Collins

PRACTICE MULTIPLE CHOICE QUESTIONS

CAPE® Physics

Peter DeFreitas

Collins

HarperCollins Publishers Ltd
The News Building
1 London Bridge Street
London SE1 9GF

First edition 2016

10 9 8 7 6 5 4 3 2

© HarperCollins *Publishers* Limited 2016

ISBN 9780-00-820509-6

Collins® is a registered trademark of HarperCollins Publishers Limited

CAPE Physics is an independent publication and has not been authorised, sponsored or otherwise approved by **CXC®**.

CAPE® is a registered trade mark of the **Caribbean Examinations Council (CXC®)**.

www.collins.co.uk/caribbeanschools

A catalogue record for this book is available from the British Library.

Typeset by QBS Learning
Printed by CPI Group (UK) Ltd, Croydon, CR0 4YY

All rights reserved. No part of this book may be reproduced, stored in a retrieval system, or transmitted in any form or by any means, electronic, mechanical, photocopying, recording or otherwise, without the prior permission in writing of the Publisher. This book is sold subject to the conditions that it shall not, by way of trade or otherwise, be lent, re-sold, hired out or otherwise circulated without the Publisher's prior consent in any form of binding or cover other than that in which it is published and without a similar condition including this condition being imposed on the subsequent purchaser.

If any copyright holders have been omitted, please contact the Publisher who will make the necessary arrangements at the first opportunity.

Author: Peter DeFreitas
Publisher: Elaine Higgleton
Commissioning Editor: Ben Gardiner
Managing Editor: Sarah Thomas
Project Manager: Alissa McWhinnie
Copy Editor: David Batty
Proofreader: Helen Bleck
Answer Checker: Jane Roth
Collation: Aidan Gill
Artwork: QBS Learning
Cover design: Kevin Robbins and Gordon MacGilp
Production: Lauren Crisp

Contents

Introduction v

Unit 1: Mechanics, Oscillations and Waves, Thermal and Mechanical Properties of Matter 1

Module 1: Mechanics 2
- **1.1.1** Physical Quantities, SI Units and Vectors 2
- **1.1.2** Forces, Statics and Linear Motion 8
- **1.1.3** Forces and Non-linear Motion 21
- **1.1.4** Forces, Momentum and Energy 27

Module 2: Oscillations and Waves 34
- **1.2.1** Harmonic Motion 34
- **1.2.2** Properties of Waves 1 42
- **1.2.3** Properties of Waves 2 52
- **1.2.4** Physics of the Ear and Eye 57

Module 3: Thermal and Mechanical Properties of Matter 62
- **1.3.1** Design and Use of Thermometers 62
- **1.3.2** Thermal Properties 65
- **1.3.3** Heat Transfer 68
- **1.3.4** The Kinetic Theory of Gases 73
- **1.3.5** First Law of Thermodynamics 78
- **1.3.6** Mechanical Properties of Materials 84

Unit 2: Electricity and Magnetism, A.C. Theory and Electronics, Atomic and Nuclear Physics 93

Module 1: Electricity and Magnetism 94
- **2.1.1** Electrical Quantities 94
- **2.1.2** Electrical Circuits 97
- **2.1.3** Electric Fields 107
- **2.1.4** Capacitors 112
- **2.1.5** Magnetic Fields and Forces 116
- **2.1.6** Electromagnetic Induction 123

Module 2: A.C. Theory and Electronics ... 128
2.2.1 Alternating Currents ... 128
2.2.2 The p-n Junction Diode and Transducers ... 131
2.2.3 Operational Amplifiers ... 138
2.2.4 Logic Gates ... 150

Module 3: Atomic and Nuclear Physics ... 157
2.3.1 Particulate Nature of Electromagnetic Radiation ... 157
2.3.2 Atomic Structure and Binding Energy ... 168
2.3.3 Radioactivity ... 174

Download answers for free at www.collins.co.uk/caribbeanschools

Introduction

Structure of the CAPE® Physics Paper 1 Examination

There are **45 questions** in the Unit 1 examination and **45 questions** in the Unit 2 examination. The duration of each examination is **1 ½ hours**. The paper is worth **40%** of your final examination mark.

The Paper 1 examinations test the following core areas of the syllabus.

Unit 1: Mechanics, Oscillations and Waves, Thermal and Mechanical Properties of Matter

Section	Number of Questions
Module 1: Mechanics	15
Module 2: Oscillations and Waves	15
Module 3: Thermal and Mechanical Properties of Matter	15
Total	45

Unit 2: Electricity and Magnetism, A.C. Theory and Electronics, Atomic and Nuclear Physics

Section	Number of Questions
Module 1: Electricity and Magnetism	15
Module 2: A.C. Theory and Electronics	15
Module 3: Atomic and Nuclear Physics	15
Total	45

The questions test two profiles, **knowledge and comprehension**, and **use of knowledge**. Questions will be presented in a variety of ways including the use of diagrams, data, graphs, prose or other stimulus material.

Each question is allocated 1 mark. You will not lose a mark if a question is answered incorrectly.

Examination Tips

General strategies for answering multiple choice questions

- Read every word of each question very carefully and make sure you understand exactly what it is asking. Even if you think that the question appears simple or straight forward there may be important information you could easily omit, especially small, but very important words such as *all* or *only*.
- When faced with a question that seems unfamiliar, re-read it very carefully. Underline or circle the key pieces of information provided. Re-read it again if necessary to make sure you are very clear as to what it is asking and that you are not misinterpreting it.
- Each question has four options, **A**, **B**, **C** and **D**, and only one is the correct answer. Look at all the options very carefully as the differences between them may be very subtle; never stop when you come across an option you think is the one required. Cross out options that you know are incorrect for certain. There may be two options that appear very similar; identify the difference between the two so you can select the correct answer.
- You have approximately 2 minutes per question. Some questions can be answered in less than 1 minute while other questions may require longer because of the reasoning or calculation involved. Do not spend too long on any one question.

- If a question appears difficult place a mark, such as an asterisk, on your answer sheet alongside the question number and return to it when you have finished answering all the other questions. Remember to carefully remove the asterisk, or other markings, from the answer sheet using a good clean eraser as soon as you have completed the question.
- Answer every question. Marks are not deducted for incorrect answers. Therefore, it is in your best interest to make an educated guess in instances where you do not know the answer. Never leave a question unanswered.
- Always ensure that you are shading the correct question number on your answer sheet. It is very easy to make a mistake, especially if you plan on returning to skipped questions.
- Some questions may ask which of the options is NOT correct or is INCORRECT. Pay close attention to this because it is easy to fail to see the words *NOT* or *INCORRECT* and answer the question incorrectly.
- Some questions may give two or more answers that could be correct and you are asked to determine which is the *BEST* or *MOST LIKELY*. You must consider each answer very carefully before making your choice because the differences between them may be very subtle.
- When a question gives three or four answers numbered **I, II** and **III** or **I, II, III** and **IV**, one or more of these answers may be correct. You will then be given four combinations as options, for example:

  (A) I only

  (B) I and II only

  (C) II and III only

  (D) I, II and III

Place a tick by all the answers that you think are correct before you decide on the final correct combination.

- Do not make any assumptions about your choice of options, just because two answers in succession have been C, it does not mean that the next one cannot be C as well.
- Try to leave time at the end of the examination to check over your answers, but never change an answer until you have thought about it again very carefully.

Strategies for the CAPE® Physics Paper 1

- A silent, non-programmable calculator is allowed in the examination. You are required to provide your own calculator. Since the different brands of calculators have unique features it is advisable to take a calculator that you are familiar with.
- Switching the calculator between radians and degrees is often required. Do not forget to change the mode to radians when necessary – particularly in topics such as simple harmonic motion, or in general waveforms such as in A.C. theory. For example, to use the formula, $y = A \sin \omega t$, your calculator should be set to radians.
- For numerical answers, ensure that you have considered the prefix to the unit. You may have calculated a value to be 0.5 kg but the answer may be presented as 500 g.
- Know your conversion of certain units that involve areas or volumes. For example, using SI base units, 2 mm^3 is 2×10^{-9} m^3.
- It is often very useful to sketch diagrams in order to obtain a combined view of several individual items. This is particularly important for questions where the effect of several forces is needed to formulate an equation.
- You must know all of the required formulas. Have these memorised long before the examination so that they will be firmly installed in your memory.

Unit 1: Mechanics, Oscillations and Waves, Thermal and Mechanical Properties of Matter

Module 1: Mechanics
1.1.1: Physical Quantities, SI Units and Vectors

1 Which of the following quantities has no dimensions?

   I. Magnification
   II. Refractive index
   III. Universal gravitational constant
   IV. Relative density

(A) I and II only
(B) I and III only
(C) II and IV only
(D) I, II and IV only

2 The molar gas constant can be expressed in terms of

(A) $kg\ s^2\ m^{-2}\ mol^{-1}$
(B) $kg\ m^2\ s^{-2}\ K^{-1}\ mol^{-1}$
(C) $kg\ m^2\ K^{-1}\ mol^{-1}$
(D) $kg\ K^{-1}\ mol^{-1}$

3 Which of the following groups is comprised only of SI base units?

(A) Gram, metre
(B) Newton, pascal
(C) Mole, ampere
(D) Kilogram, newton

Item 4 refers to the following quantities together with their SI base units.

$E: \text{kg m}^2 \text{ s}^{-2}$ $F: \text{kg m s}^{-2}$ $G: \text{kg s}^{-1}$

4 The quantity represented by $\dfrac{EG}{F}$ is

(A) displacement
(B) pressure
(C) momentum
(D) velocity

5 F is the force acting on an area, A, and doing an amount of work, W, on an object of density ρ. The expression $\dfrac{F^2}{WA\rho}$ has the unit of

(A) acceleration
(B) speed
(C) displacement
(D) pressure

6 The SI unit of pressure may be expressed as

(A) $\text{kg s}^{-2} \text{ m}^{-2}$
(B) $\text{kg s}^{-2} \text{ m}^{-1}$
(C) $\text{kg}^2 \text{ s}^{-2} \text{ m}^{-1}$
(D) kg m^{-2}

7 Given that P is 20.0 MJ and Q is 5.0 mJ, then the value of P is

(A) $4.0 \times 10^6 \, Q$
(B) $2.5 \times 10^8 \, Q$
(C) $4.0 \times 10^9 \, Q$
(D) $2.5 \times 10^9 \, Q$

1.1.1: Physical Quantities, SI Units and Vectors (cont.)

Items **8–9** refer to Tia and her baby, mentioned below.

Tia stands on her bathroom scale which registers her mass as (42 ± 1) kg. She then lifts her baby and the new reading on the scale is (46 ± 1) kg.

8 The mass of her baby is

(A) (4 ± 1) kg
(B) (3 ± 1) kg
(C) (4 ± 2) kg
(D) (5 ± 1) kg

Ⓐ
Ⓑ
Ⓒ
Ⓓ

9 The percentage error (uncertainty) in the measurement of the baby's mass is

(A) 50%
(B) 25%
(C) 2.5%
(D) 5.0%

Ⓐ
Ⓑ
Ⓒ
Ⓓ

10 The diameter, d, of a small circular play area is 5.0 m ± 0.1 m. The area, A, is calculated using the equation $A = \pi \frac{d^2}{4}$. The percentage uncertainty in the result is

(A) 8%
(B) 4%
(C) 1%
(D) 2%

Ⓐ
Ⓑ
Ⓒ
Ⓓ

11 In which of the following cases will systematic error be reduced?

(A) Determining the force constant of a spring from the gradient of a force–extension graph produced from the plot of several pairs of readings of force and extension as the spring is loaded.

(B) Setting the pointer of an ammeter to read exactly zero when no current flows through it.

(C) Finding the mean time of a 100 m race using the values measured by three persons, each with a stop watch.

(D) Determining the length of a drinking straw by finding the average value of three measurements of its length.

12 Exactly 210 cm^3 of water of density 1.0 g cm^{-3} is placed in a measuring cylinder. A small object of mass 8.0 g floats when inserted into the cylinder. Determine the new volume reading.

(A) 218 cm^3

(B) 226 cm^3

(C) 210 cm^3

(D) 215 cm^3

13 The number of moles of uranium-235 atoms in 0.047 kg of uranium-235 is

(A) 2.0

(B) 0.2

(C) 5.0

(D) 0.5

14 The molar mass of carbon is 12 g. If the density of diamond is 3500 kg m^{-3}, then the number of carbon atoms in 2.0 cm^3 of diamond is

(A) 1.0×10^{23}

(B) 3.5×10^{17}

(C) 1.0×10^{24}

(D) 3.5×10^{23}

1.1.1: Physical Quantities, SI Units and Vectors (cont.)

15 Force vectors *P* and *Q* have magnitudes of 30 N and 50 N respectively. Which of the following is NOT a possible answer for the magnitude of the resultant of these vectors?

(A) 15 N

(B) 30 N

(C) 45 N

(D) 60 N

Item **16** refers to the following diagram, which shows three coplanar forces in equilibrium.

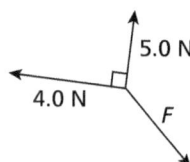

16 The magnitude of *F* is

(A) 9.0 N

(B) 41 N

(C) 6.4 N

(D) 3.0 N

Item **17** refers to the vectors *M* and *L*.

17 Which figure below BEST represents *N*, the addition of *L* and *M*?

(A)

(B)

(C)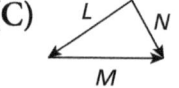

(D)

18 $k = MN$ where k is a constant and M and N are variables. A graph of M against N would look like

(A)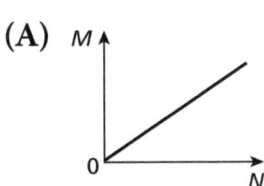

(B)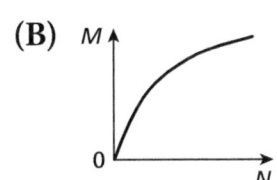

(C)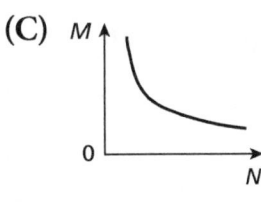

(D)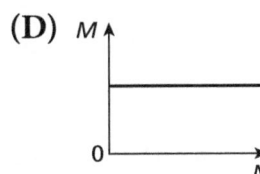

19 Graphs like the ones below can be used to interpret the accuracy and precision of measurements. Several values of the diameter, d, of a pin were taken and the number of occurrences, N, of particular values of d was plotted against d. The actual diameter of the pin is d_0.

The graph that indicates poor precision and high accuracy is

(A)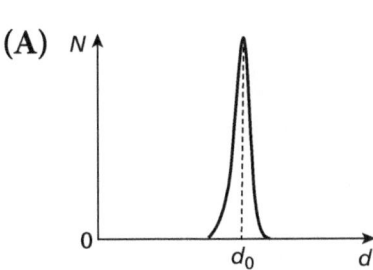

(B)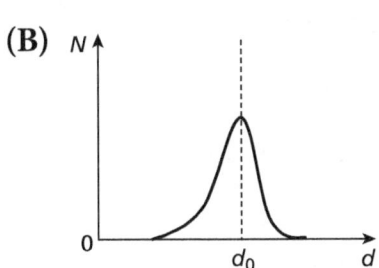

(C)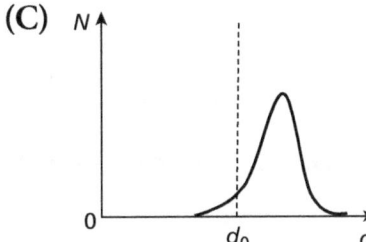

(D)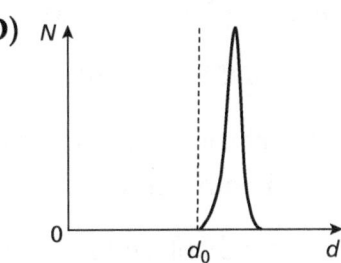

1.1.1: Physical Quantities, SI Units and Vectors (cont.)

20 The diameter of a small sphere needs to be found. Josh has two of these spheres similar in shape and size, and arranges them as shown alongside a metre rule.

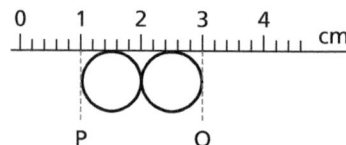

It is estimated that the uncertainty of the reading at each point is ± 0.2 cm.

The mean diameter of one of the spheres together with its uncertainty is best represented as

(A) (1.0 ± 0.1) cm

(B) (1.0 ± 0.2) cm

(C) (1.0 ± 0.4) cm

(D) (1.00 ± 0.20) cm

1.1.2: Forces, Statics and Linear Motion

1 Which of the following is NOT true for an object floating in water?

(A) The upthrust is equal in magnitude to the weight of water displaced.

(B) The weight of water displaced is equal to the weight of the portion of the object that is submerged.

(C) The weight of water displaced is equal to the weight of the object.

(D) The upthrust is equal in magnitude to the weight of the object.

2 A balloon filled with helium has a volume of 25 m³ and is released from rest in air of density 1.3 kg m⁻³. Determine the initial resultant force on it if the weight of the balloon and its contents is 300 N.

(A) 319 N upward

(B) 19 N upward

(C) 19 N downward

(D) 319 N downward

Item 3 refers to the torque produced by a couple of two equal forces, F, which act in opposite directions on a rigid body.

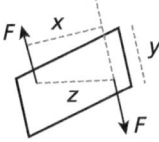

3 The torque of the couple is

(A) Fx

(B) Fy

(C) $2Fz$

(D) $2Fx$

4 The diagram below shows a rod kept stationary under the action of coplanar forces, L, M, N.

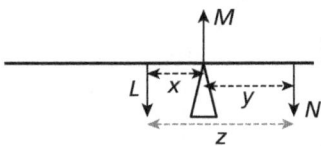

Which of the following is NOT true?

(A) $M = L + N$

(B) $My = Lz$

(C) $Nz = M(z - y)$

(D) $Mx = Ny$

1.1.2: Forces, Statics and Linear Motion (cont.)

Item 5 shows an elevation of a uniform shutter of weight 25 N hinged at its upper edge. The shutter is kept open in a stationary position by a breeze. A thin horizontal bar of weight 10 N is hooked onto the lower edge of the shutter.

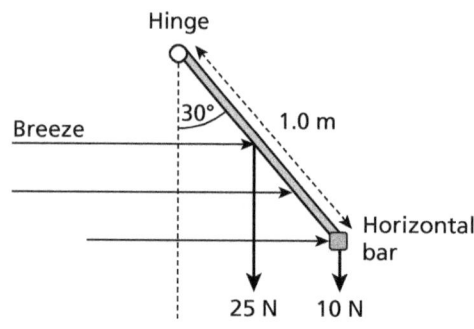

5 The torque of the breeze about the hinge is

(A) 23 N m

(B) 26 N m

(C) 11 N m

(D) 18 N m

Ⓐ
Ⓑ
Ⓒ
Ⓓ

6 The small sphere of weight, W, is suspended from point, P and is held in the position shown when the force F is applied. Which of the following is true?

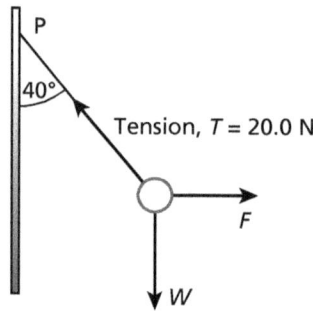

(A) $W = 15.3$ N

(B) $F = 15.3$ N

(C) $W = T \sin 40$

(D) $F = T \cos 40$

Ⓐ
Ⓑ
Ⓒ
Ⓒ

10

7 A frictional force f acts on a block of mass m as it slides down the incline shown in the diagram below.

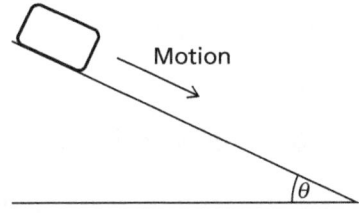

The acceleration of the block is

(A) $g \sin \theta$

(B) $g \sin \theta - \dfrac{f}{m}$

(C) $g \cos \theta - \dfrac{f}{m}$

(D) $\dfrac{f}{m}$

8 A body falls at terminal velocity. Which of the following is true?

(A) Gravitational potential energy transforms to thermal energy.

(B) The net force is in the direction of motion.

(C) The weight of the body is greater than the sum of the drag force and the upthrust.

(D) The acceleration of the body is equal to the acceleration due to gravity.

9 A fixed force F causes a body of mass m to accelerate from rest through a fixed distance, d. If the mass m is varied it is found that the velocity attained, v, is proportional to

(A) $\dfrac{1}{\sqrt{m}}$

(B) Fm^2

(C) $\sqrt{md}$

(D) $\dfrac{mF}{d}$

1.1.2: Forces, Statics and Linear Motion (cont.)

10 A spider falling from the ceiling quickly comes to a terminal velocity because

(A) the resultant downward force is reduced to zero since the viscous drag force increases until it is equal to the weight of the spider.

(B) the weight of the air displaced by the spider is equal to its own weight.

(C) the buoyancy force plus the viscous drag force are equal in magnitude to the weight of the spider.

(D) the weight of the spider is equal in magnitude to the buoyancy force on it.

11 A parcel of mass 50 kg falls from a plane and acquires a terminal velocity of 5.0 m s^{-1}. The resultant force on the parcel is then

(A) 250 N

(B) 10 N

(C) 500 N

(D) 0 N

12 Sarah, of mass 45 kg, stands in a helicopter that accelerates vertically downward at 4.0 m s^{-2}. The force exerted by the floor onto her feet is

(A) 440 N

(B) 180 N

(C) 620 N

(D) 260 N

13 A block of mass *m* is connected to one of mass 1.4 kg by a light helical spring as shown in the diagram. The blocks rest on a smooth frictionless surface. They are pulled apart so that the spring stretches and they are then released.

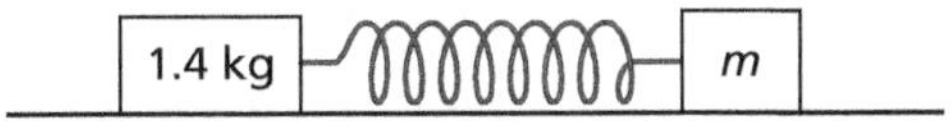

$$\boxed{1.4 \text{ kg}} \text{—}\text{000000}\text{—} \boxed{m}$$

The acceleration of the mass, *m*, is 3.5 times the acceleration of the other block. The mass, *m*, is

(A) 2.5 kg   Ⓐ

(B) 400 g   Ⓑ

(C) 2500 g   Ⓒ

(D) 40 g   Ⓓ

14 A spaceship travels through outer space expelling its burnt fuel at 7.5×10^3 m s^{-1} relative to itself. It uses fuel at a rate of 200 kg s^{-1}. When the acceleration of the spaceship is 15 m s^{-2}, its mass is

(A) 5.0×10^6 kg   Ⓐ

(B) 3.8×10^6 kg   Ⓑ

(C) 1.0×10^5 kg   Ⓒ

(D) 7.5×10^4 kg   Ⓓ

15 An object of mass 4.0 kg is pulled across a horizontal surface by a horizontal force, *F*. If its acceleration is 2.0 m s^{-2} when the frictional force is 6.0 N, the pulling force *F* is

(A) 7.5 N   Ⓐ

(B) 6.0 N   Ⓑ

(C) 8.0 N   Ⓒ

(D) 14 N   Ⓓ

1.1.2: Forces, Statics and Linear Motion (cont.)

16 An object of mass m is acted on by a single and constant force, F. Its acceleration

(A) increases.

(B) is uniform.

(C) depends on the distance travelled at any particular time.

(D) decreases.

Item 17 refers to the motion of the blocks shown in the diagram below. The slopes are of the same gradient but N has a greater mass than M. The blocks are connected by a light inextensible string and the slopes and pulley are smooth and frictionless.

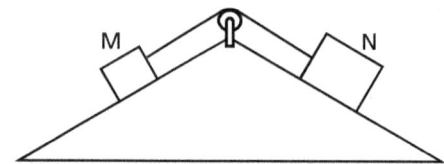

17 Which of the following is true as one block slides up the incline and the other slides down?

(A) The acceleration of N is greater than that of M.

(B) The acceleration of N increases faster than that of M.

(C) The accelerations of N and M increase at the same rate.

(D) The accelerations of N and M are the same and do not change.

Item 18 shows a block as it slides down a rough slope at constant velocity. It then continues along a horizontal smooth, frictionless surface until it reaches an edge E and free falls.

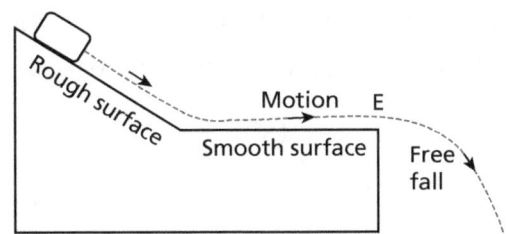

18. Which of the following force–time graphs BEST illustrates how the net force on the object varies with time?

(A)

(C)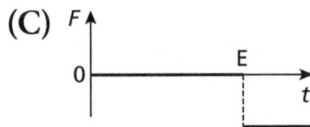

(B)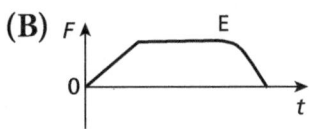

(D)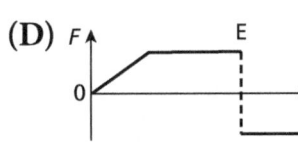

1.1.2: Forces, Statics and Linear Motion (cont.)

Items 19–21 refer to the graphs below, which represent the motion of a body, initially at rest, falling through a thick viscous fluid. Displacement is measured from the starting point.

(W)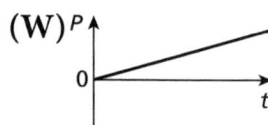

(Y)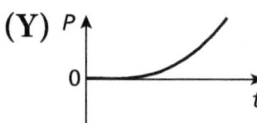

(X)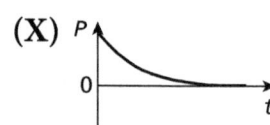

(Z)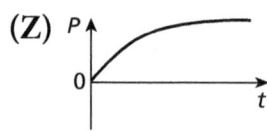

19 If P represents displacement, which graph BEST describes the motion?

(A) Z

(B) Y

(C) W

(D) X

20 If P represents velocity, which graph BEST illustrates the motion?

(A) X

(B) W

(C) Y

(D) Z

21 If P represents acceleration, which graph BEST illustrates the motion?

(A) Y

(B) Z

(C) W

(D) X

Item **22** refers to the graph below.

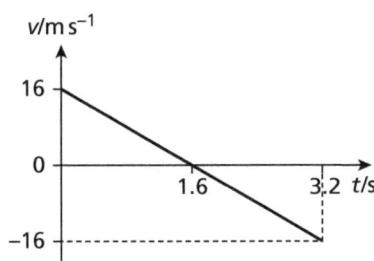

22 The graph shows the motion of a rubber ball. Which of the following events can it be representing?

(A) Falling through a distance of more than 25 m and then rebounding. Ⓐ

(B) Rolling down an incline with an acceleration of magnitude 10 m s^{-2}. Ⓑ

(C) Rising about 13 m vertically into the air and then falling. Ⓒ

(D) Falling with an acceleration of 10 m s^{-2} and then rebounding. Ⓓ

Item **23** refers to the following graph, which represents the motion of an object falling to the ground and then rebounding.

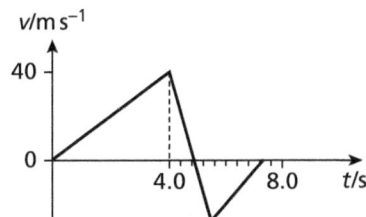

23 The height through which the object fell initially is

(A) 160 m Ⓐ

(B) 80 m Ⓑ

(C) 88 m Ⓒ

(D) 176 m Ⓓ

1.1.2: Forces, Statics and Linear Motion (cont.)

24 Erica releases her toy soldier from the roof deck of a tall building and observes it as it falls freely for just about 12 m before its parachute is engaged. Which of the following graphs BEST illustrates how its acceleration varies with time for the first few seconds of its motion?

(A)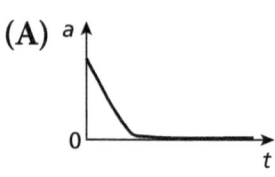

(C)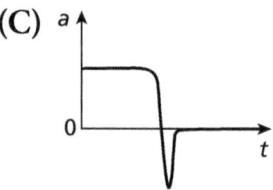

(B)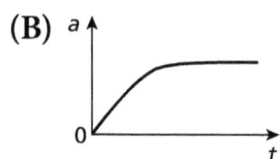

(D)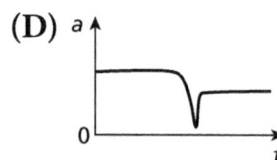

25 A small child holds his toy gun at a height of 1.0 m above the ground and shoots a dart which leaves the nozzle horizontally at a velocity of 5.0 m s^{-1}. Neglecting air resistance, the time taken for the dart to reach the floor is

(A) 0.90 s

(B) 0.45 s

(C) 1.8 s

(D) 0.51 s

26 The instantaneous velocity of an object is zero. Its acceleration

(A) is zero.

(B) may be non-zero.

(C) must be increasing or decreasing.

(D) is constant.

27 A man of mass m stands in an elevator that moves upward with acceleration a. If the acceleration due to gravity is g, the force which he exerts on the floor is

(A) mg

(B) $m(a + g)$

(C) $m(a - g)$

(D) $m(g - a)$

28 Blocks P and Q are moving along together. P is 4 times the mass of Q and is acted on by a force of 15 N as shown.

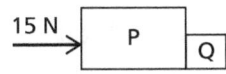

The force exerted by Q onto P is

(A) 3 N

(B) 15 N

(C) 75 N

(D) 60 N

29 A cannon ball leaves its cannon at an angle of 30° above the horizontal. Which of the following graphs represents how the vertical component v_y of its velocity varies with time t as it travels through the air and then strikes the ground on the same level as it left the cannon?

(A) (B) (C) (D)

Item 30 refers to the following displacement–time graph.

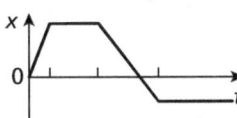

30 The velocity–time graph representing the motion shown by the displacement–time graph above is

(A) (B) (C) (D)

1.1.2: Forces, Statics and Linear Motion (cont.)

31 The following graph illustrates how the velocity v of a rubber ball changes with time t as it rises into the air, falls to the ground, and then rebounds.

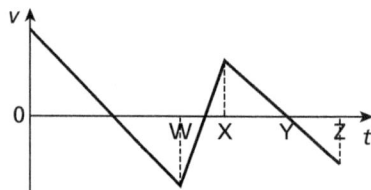

The time on reaching the highest point after rebounding is

(A) W
(B) X
(C) Y
(D) Z

Item 32 refers to the following graph, which shows the variation of acceleration a with time t for a particular situation.

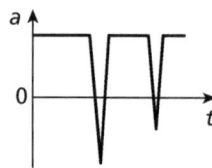

32 The graph could represent a ball tossed into the air, falling to the floor and rebounding twice

(A) to the same height, each time making an elastic collision.
(B) to the same height, each time making an inelastic collision.
(C) to a lesser height, each time making an inelastic collision.
(D) to a lesser height, each time making an elastic collision.

Item 33 refers to the following velocity–time graph.

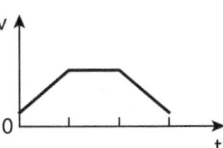

33 Which of the following displacement–time graphs BEST represents the motion of the velocity–time graph?

(A)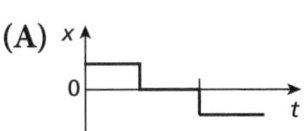

(C)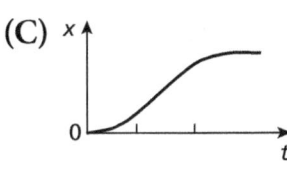

(B)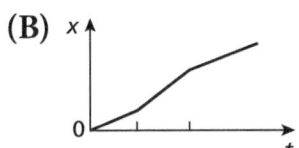

(D)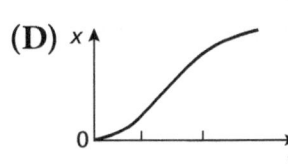

Ⓐ
Ⓑ
Ⓒ
Ⓓ

1.1.3: Forces and Non-linear Motion

1 An object is thrown through the air to the top of a ledge as shown.

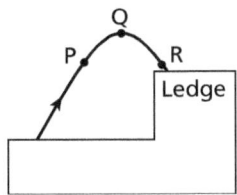

Which of the following is NOT true of the motion at P, Q and R?

(A) The vertical component of the velocity changes but the horizontal component of the velocity is constant.

(B) The vertical and horizontal components of the acceleration are constant and zero respectively.

(C) The vertical component of the acceleration at Q is zero.

(D) The vertical component of the acceleration at P is in the same direction as the vertical component of the acceleration at R.

Ⓐ
Ⓑ
Ⓒ
Ⓓ

21

1.1.3: Forces and Non-linear Motion (cont.)

2 Which of the following is NOT true of a geostationary satellite?

(A) Its orbital speed is independent of its mass.

(B) It does not accelerate.

(C) Its angular speed is the same as that of the rotation of the Earth.

(D) The square of its angular speed is inversely proportional to the cube of the radius of its orbit.

3 The second hand of a watch is of length 2.00 cm. Its angular speed is

(A) 5.25×10^{-2} rad s^{-1}

(B) 1.05×10^{-1} rad s^{-1}

(C) 3.14×10^{2} rad s^{-1}

(D) 60.0 rad s^{-1}

4 Using the acceleration due to gravity at the surface of the Earth as 9.8 m s^{-2}, what would be the acceleration due to gravity at the surface of a planet of mass twice that of the Earth and of radius four times that of the Earth?

(A) 2.5 m s^{-2}

(B) 5.0 m s^{-2}

(C) 7.5 m s^{-2}

(D) 1.2 m s^{-2}

5 Planet X is of density ρ and has a value of acceleration due to gravity of g at its surface. Planet Y is of density twice that of X and the acceleration due to gravity at its surface is $4g$. Assuming the planets are uniform spheres, if the radius of Y is r, what is the radius of X?

(A) $2r$

(B) $\frac{r}{2}$

(C) $\frac{r}{8}$

(D) $8r$

6 Which of the following statements is/are true for a uniform gravitational field?

  I. The direction of the field is the same as the direction moved by a small mass immersed in it.

  II. The field strength at all points within the field is the same.

  III. Since the field is uniform the same force is exerted on any particle placed in it.

(A) I only

(B) I and II only

(C) II only

(D) II and III only

7 The resultant force on a person at the highest point of his orbit in a ride on a Ferris wheel is 200 N. An upward force of 200 N is suddenly applied to him from the seat. His immediate resulting motion is

(A) to an orbit of larger radius.

(B) along the tangent of the orbit at the point where he leaves the orbit.

(C) to an orbit of smaller radius.

(D) vertically up.

8 Cars X and Y speed around the same corner. The frictional force on their tyres providing the circular motion is the same. Which of the following must be equal for X and Y?

(A) Mass

(B) Kinetic energy

(C) Angular speed

(D) Acceleration

1.1.3: Forces and Non-linear Motion (cont.)

9 A stone tied to the end of a string of length 2.0 m is whirled in a circle at constant speed v. What is the speed if it moves through an angle of 1.57 radians, in a time of 0.40 seconds?

(A) 1.25 m s^{-1}

(B) 0.63 m s^{-1}

(C) 7.9 m s^{-1}

(D) 49 m s^{-1}

Ⓐ Ⓑ Ⓒ Ⓓ

10 An object of mass m moves with uniform circular motion of period T in an orbit of radius r. Which of the following is true of the mass?

(A) The net force on it is zero.

(B) Its velocity is constant.

(C) Its acceleration is $\dfrac{4\pi r^2}{T^2}$.

(D) No work is done on it although there is a force on it and it is moving through a distance.

Ⓐ Ⓑ Ⓒ Ⓓ

11 An object of mass, 2.0 g, moves with uniform circular motion. Its radial acceleration is 4.0 m s^{-2} and the radius of its orbit is 1.2 m. The period of its motion is

(A) 0.30 s

(B) 1.9 s

(C) 11.8 s

(D) 3.4 s

Ⓐ Ⓑ Ⓒ Ⓓ

12 A boy of mass 50 kg rides on a Ferris wheel which rotates so rapidly that he just becomes weightless at the highest point. The diameter of the wheel is 10 m. What is the reaction force, R, from the seat and the speed, v, when at the lowest point in this motion?

(A) $R = 490$ N $v = 9.9$ m s^{-1} Ⓐ

(B) $R = 980$ N $v = 9.9$ m s^{-1} Ⓑ

(C) $R = 490$ N $v = 19.8$ m s^{-1} Ⓒ

(D) $R = 980$ N $v = 19.8$ m s^{-1} Ⓓ

13 A car of mass m travelling at constant speed v moves over the brim of a hill of radius r. What is the reaction force R on the car from the road if it does not leave the road surface?

(A) $R = \dfrac{mv^2}{r} - mg$ (C) $R = mg - \dfrac{mv^2}{r}$

(B) $R = \dfrac{mv^2}{r} + mg$ (D) $R = \dfrac{mv^2}{r}$

Ⓐ Ⓑ Ⓒ Ⓓ

14 The maximum safe speed for a car to round a particular corner on a level track is 25 m s^{-1}. An oil mixture spills on the road and reduces the frictional force exerted on the tyres by 80%. To safely drive round the corner the driver must reduce his speed to

(A) $5\sqrt{5}$ m s^{-1} Ⓐ

(B) 5 m s^{-1} Ⓑ

(C) 25 m s^{-1} Ⓒ

(D) $\sqrt{5}$ m s^{-1} Ⓓ

15 A stone of mass 500 g attached to a piece of string, is whirled at constant speed v in a vertical circle of radius 2.0 m. What is the speed v if the tension in the string when the stone is at the lowest point is 9.0 N?

(A) 36 m s^{-1} Ⓐ

(B) 6.0 m s^{-1} Ⓑ

(C) 16 m s^{-1} Ⓒ

(D) 4.0 m s^{-1} Ⓓ

1.1.3: Forces and Non-linear Motion (cont.)

16 A small spaceship of mass 500 kg on Earth is placed in orbit at a height above the surface equal to 1.5 times the diameter of the planet. What is the gravitational force on the spaceship keeping it in orbit?

(A) 2400 N

(B) 1200 N

(C) 307 N

(D) 150 N

17 Satellites X and Y orbit a planet. Satellite X orbits just above the surface at radius R_x and the speeds of X and Y are, respectively, 4.0 m s^{-1} and 1.0 m s^{-1}. The orbit radius R_y of the other satellite is

(A) $16R_x$

(B) $4R_x$

(C) $2R_x$

(D) $\dfrac{R_x}{16}$

18 A satellite of mass m is at a distance r above the surface of a planet of radius R. The acceleration due to gravity at the surface of the planet is g. What is the centripetal force on the satellite?

(A) $mg\dfrac{R^2}{r^2}$

(B) $mg\dfrac{(R+r)^2}{(r)^2}$

(C) $mg\dfrac{R^2}{(R+r)^2}$

(D) $mg\dfrac{r^2}{(R+r)^2}$

1.1.4: Forces, Momentum and Energy

1 Which of the following is/are true for a stone whirled on a string in a vertical circle with uniform circular motion?

   I. The velocity and acceleration are constant.

   II. The magnitude of the resultant force on the stone as well as the kinetic energy of the stone remain constant.

   III. The tension in the string changes.

(A) I and II only

(B) I and III only

(C) II only

(D) II and III only

2 Which of the following graphs BEST represents the variation of kinetic energy E with time t for a small metal sphere falling from rest through a viscous fluid and acquiring a terminal velocity?

(A)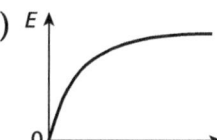

(B)

(C)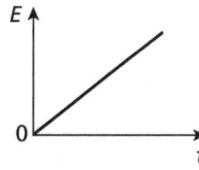

(D)

1.1.4: Forces, Momentum and Energy (cont.)

3 A football is kicked from the ground of a football field at an initial angle of about 60° above the horizontal. It travels through the air and lands neatly between the turf and the foot of the awaiting striker. Neglecting air resistance, which graph BEST represents the variation of the kinetic energy E of the ball with the VERTICAL distance x it travels?

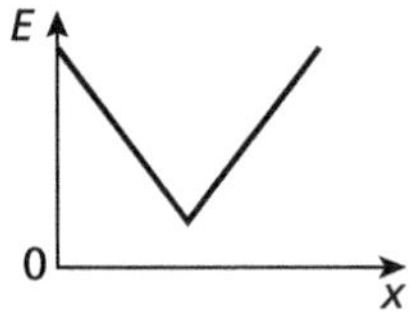

(A), (B), (C), (D)

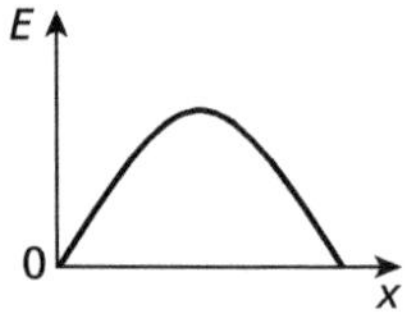

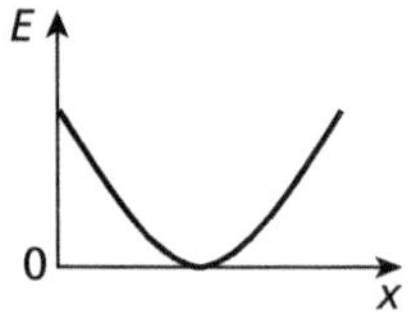

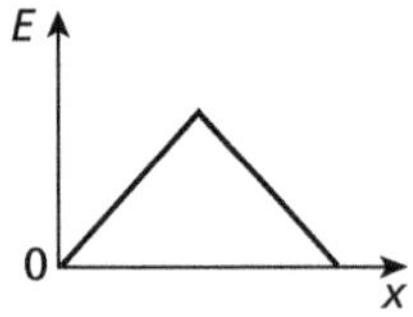

Ⓐ Ⓑ Ⓒ Ⓓ

4 An object is acted on by a single constant force. Which of the following graphs BEST illustrates the relation between its momentum p and the distance D it travels?

(A), (B), (C), (D)

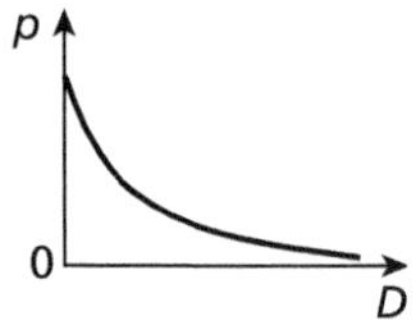

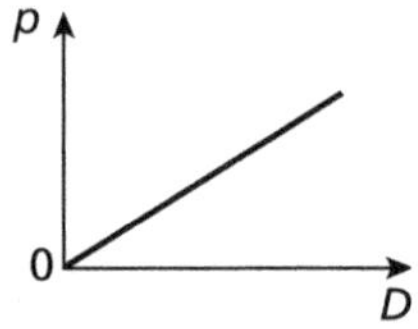

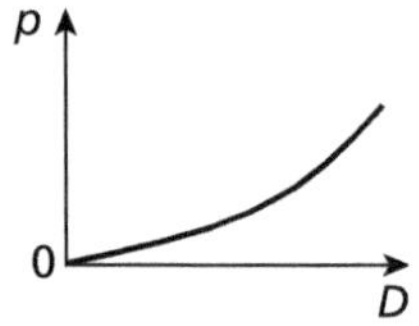

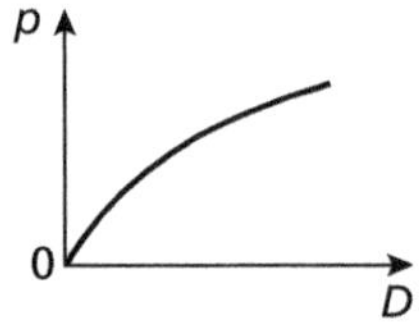

Ⓐ Ⓑ Ⓒ Ⓓ

5 Which of the following statements is/are true for a collision between two objects, A and B?

   I. If the collision is elastic, the total momentum as well as the total kinetic energy are conserved.

   II. If the objects stick together, the collision is inelastic and the total momentum is conserved.

   III. The momentum of A immediately before and immediately after the collision is the same.

(A) I only

(B) I and II only

(C) I and III only

(D) II and III only

Items **6–7** refer to the following situation.

A resultant force of 20 N acts on an object of mass 4.0 kg for a period of 0.50 s. The object is initially moving at 2.0 m s^{-1} in the direction of the resultant force.

6 The impulse on the object is

(A) 5.0 N s

(B) 10 kg m s^{-1}

(C) 40 N s

(D) 2.0 kg m s^{-1}

7 The velocity acquired at the end of the period of 0.50 s is

(A) 5.0 m s^{-1}

(B) 40 m s^{-1}

(C) 4.5 m s^{-1}

(D) 9.0 m s^{-1}

1.1.4: Forces, Momentum and Energy (cont.)

8 A block is acted on by a resultant force which is directed along a straight line and varies with time as shown in the following graph.

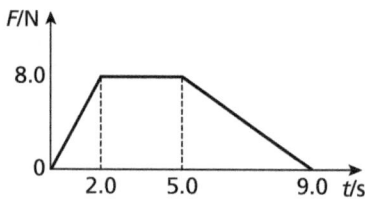

If the block is initially stationary and acquires a velocity of 4.0 m s^{-1}, its mass must be

(A) 48 kg

(B) 12 kg

(C) 2.0 kg

(D) 1.0 kg

9 Two objects, each of mass m, travel in opposite directions along the same line of motion and collide elastically. If the initial speed of one is twice that of the other, which of the following statements is/are true?

I. After the collision the total momentum is $\frac{3}{2}mv$ and the total kinetic energy is $\frac{5}{8}mv^2$.

II. The total momentum and the total kinetic energy remain constant.

III. The objects have a common velocity immediately after the collision.

(A) I and II only

(B) II only

(C) II and III only

(D) I, II and III

10 A body of mass m moving north at a speed u collides head-on with another of mass $4m$ moving at a speed $2u$ in the opposite direction. After the collision they have a common velocity v.

This common velocity is

(A) $\frac{9}{5}u$ south

(B) $\frac{7}{5}u$ south

(C) $\frac{9}{5}u$ north

(D) $\frac{7}{5}u$ north

Items **11–12** refer to the following situation.

A net force of 50 N is applied to a body of mass 4.0 kg, causing it to accelerate from rest to a velocity of 2.0 m s^{-1}.

11 The time taken for this to occur is

(A) 12.5 s

(B) 25 s

(C) 0.16 s

(D) 8.0 s

12 The rate of change of momentum is

(A) 25 kg m s^{-1}

(B) 8.0 kg m s^{-2}

(C) 50 N

(D) 8.0 N

1.1.4: Forces, Momentum and Energy (cont.)

13 A small single-seat car of mass 400 kg is coasting at a constant speed of 20 m s^{-1} along a straight, level road with a driver of mass 100 kg. The driver ejects with a vertical velocity of 5.0 m s^{-1} due to a vertical force from his seat. The new velocity of the car will be

(A) 25 m s^{-1}

(B) 20 m s^{-1}

(C) 24 m s^{-1}

(D) 5.0 m s^{-1}

14 Skaters X and Y of the same mass, travelling along lines perpendicular to each other at speeds of 2.0 m s^{-1} and 4.0 m s^{-1} respectively, collide and become entangled. The angle subtended between their new path and the path of X initially is

(A) 27°

(B) 76°

(C) 14°

(D) 63°

15 Stones, X and Y, are each tied to the ends of strings of the same length. The stones are whirled in circular motion at constant speed. X has half the mass of Y but travels at four times its speed. What can be said of the kinetic energies, E_{kX} and E_{kY}, of the stones?

(A) $E_{kX} = E_{kY}$

(B) $E_{kX} = 2E_{kY}$

(C) $E_{kX} = 4E_{kY}$

(D) $E_{kX} = 8E_{kY}$

16 A small toy boat of mass 400 g uses energy at a rate of 500 J s^{-1} as it travels across a pond at a constant velocity of 2.0 m s^{-1}. The driving force on the boat is

(A) 625 N

(B) 250 N

(C) 100 N

(D) 125 N

17 A machine uses 2.0 kJ of energy from its fuel to do a job. The energy wasted in the process is only due to a frictional force of 40 N whose point of application moves through a distance of 30 m in the direction of the force. The efficiency of the machine is

(A) 40%

(B) 60%

(C) 0.040

(D) 0.60

18 A toy car of mass m and power P accelerates from rest for a period of t seconds. The maximum velocity acquired at the end of this time is:

(A) $\dfrac{2Pt}{m}$

(B) $\sqrt{\dfrac{m}{2Pt}}$

(C) $\dfrac{m}{2Pt}$

(D) $\sqrt{\dfrac{2Pt}{m}}$

19 A tension T is produced on extending the spring of a spring gun through a distance of 10 cm. When the tension is released it imparts a maximum velocity of 6.0 m s^{-1} to a shot of mass 5.0 g. What is the value of T?

(A) 1.8 N

(B) 3.6 N

(C) 18 N

(D) 36 N

Module 2: Oscillations and Waves
1.2.1: Harmonic Motion

1 The relation between displacement s and time t for a small bob demonstrating simple harmonic motion is given as $s = 4.0 \sin 0.50\pi t$ where s is measured in mm and t in s.

Which of the following gives the amplitude A and frequency f of the vibration?

(A) $A = 0.50$ mm, $f = 4.0\pi$ Hz

(B) $A = 0.50$ mm, $f = 4.0$ Hz

(C) $A = 4.0$ mm, $f = 0.50$ Hz

(D) $A = 4.0$ mm, $f = 0.25$ Hz

2 Which of the following is constant when a particle oscillates with simple harmonic motion?

(A) Kinetic energy

(B) Velocity

(C) Frequency

(D) Acceleration

Items 3–5 refer to the following diagram.

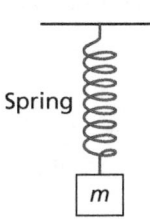

3 The period of oscillation is T_x when the mass m is pulled down a small amount and then released. If the mass is increased to $4m$, the new period becomes T_y. What is the value of the ratio $\dfrac{T_x}{T_y}$?

(A) $\dfrac{1}{2}$

(B) 2

(C) 4

(D) $\dfrac{1}{4}$

Ⓐ
Ⓑ
Ⓒ
Ⓓ

4 If the force constant of the spring is k, and the amplitude of the oscillation is A, then the values of the maximum velocity v_{max} and maximum acceleration a_{max} of the mass m are

(A) $v_{max} = \dfrac{k}{m} A \qquad a_{max} = \dfrac{k^2}{m^2} A$

(B) $v_{max} = \dfrac{m}{k} A \qquad a_{max} = \dfrac{m^2}{k^2} A$

(C) $v_{max} = \sqrt{\dfrac{k}{m}} A \qquad a_{max} = \dfrac{k}{m} A$

(D) $v_{max} = \sqrt{\dfrac{m}{k}} A \qquad a_{max} = \dfrac{m}{k} A$

Ⓐ
Ⓑ
Ⓒ
Ⓓ

5 A mass hanging at rest from the spring stretches it by a distance of 4.0 cm. It is then pulled down a further 2.0 cm and released.

What is the maximum velocity v_{max} and the maximum acceleration a_{max} attained in its oscillation if it vibrates with an angular frequency of 4π radians s^{-1}?

(A) $v_{max} = 0.32\pi^2$ m s^{-1} $\qquad a_{max} = 0.32\pi^2$ m s^{-2}

(B) $v_{max} = 0.32\pi$ m s^{-1} $\qquad a_{max} = 0.64\pi^2$ m s^{-2}

(C) $v_{max} = 0.080\pi$ m s^{-1} $\qquad a_{max} = 0.32\pi^2$ m s^{-2}

(D) $v_{max} = 0.080\pi^2$ m s^{-1} $\qquad a_{max} = 0.16\pi$ m s^{-2}

Ⓐ
Ⓑ
Ⓒ
Ⓓ

1.2.1: Harmonic Motion (cont.)

6 A body oscillating with simple harmonic motion has an amplitude of 1.2 m and a period of 2.0 s. What is its maximum velocity?

(A) 1.2π m s^{-1}

(B) 0.60 m s^{-1}

(C) 3.3π m s^{-1}

(D) 3.3 m s^{-1}

7 Which of the following is/are true for a body moving in simple harmonic motion?

  I. The acceleration varies and is always directed to a fixed point.

  II. The maximum velocity occurs at the centre of the oscillation and is always directed to a fixed point.

  III. The acceleration a at displacement x is given by the equation $a = -\omega^2 x$ where ω is the angular frequency.

(A) I only

(B) I and II only

(C) I and III only

(D) III only

8 Two similar springs are connected to the mass as shown in the following diagram. The mass m is displaced a distance A to the left and then released. If the acceleration a at any time is given by $a = -\frac{2ke}{m}$ where e is the instantaneous displacement, what is the sum of the kinetic and potential energies of the mass at any time?

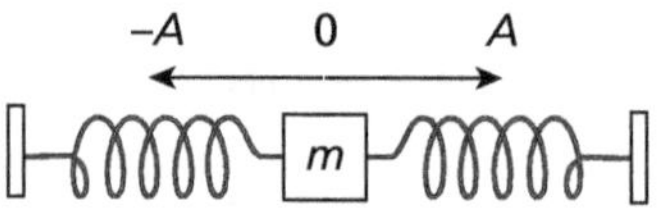

(A) $\frac{k}{m}A^2$

(B) kA^2

(C) $\frac{1}{2}\frac{k}{m}A^2$

(D) $2\frac{k}{m}e^2$

9 A pendulum has a period of T_x on planet X where the acceleration due to gravity is g_x and of T_y on planet Y. What is the acceleration due to gravity on planet Y?

(A) $g_x \dfrac{T_y^2}{T_x^2}$

(B) $g_x \dfrac{T_x}{T_y}$

(C) $g_x \dfrac{T_y}{T_x}$

(D) $g_x \dfrac{T_x^2}{T_y^2}$

Ⓐ Ⓑ Ⓒ Ⓓ

Items **10–11** refer to the following situation.

A pendulum makes 20 oscillations in 4.0 s on Earth and 40 oscillations in 4.0 s on planet X.

10 What is the acceleration due to gravity on planet X?

(A) 19.6 m s^{-2}

(B) 39 m s^{-2}

(C) 4.9 m s^{-2}

(D) 0.61 m s^{-2}

Ⓐ Ⓑ Ⓒ Ⓓ

11 A mass hung onto the lower end of a vertical spring oscillates vertically with the same period as the pendulum does on Earth.

What is the period of the mass-spring system if carried to planet X?

(A) 5.0 s

(B) 10.0 s

(C) 0.10 s

(D) 0.20 s

Ⓐ Ⓑ Ⓒ Ⓓ

12 A mass hanging from the lower end of a vertical spring oscillates with simple harmonic motion. Which of the graphs shown illustrates how its velocity v varies with time t for one oscillation, if timing is started when the mass passes through its mean position?

(A)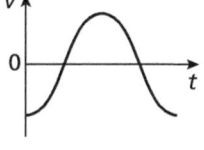

(C)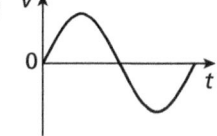

(B)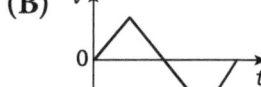

(D)

Ⓐ Ⓑ Ⓒ Ⓓ

1.1.5: Harmonic Motion (cont.)

13 Which of the following graphs does NOT illustrate simple harmonic motion?

(F = resultant force, a = acceleration, v = velocity, x = displacement from mean position, E_k = kinetic energy, E_p = potential energy)

(A)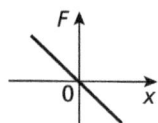

(C)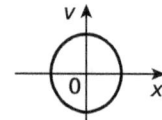

(B)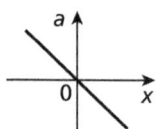

(D)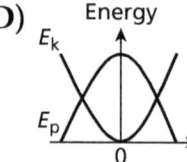

Ⓐ Ⓑ Ⓒ Ⓓ

Item 14 refers to the following graph, which shows how the amplitude A of a child's swing varies with the driving frequency f of the force she exerts on it. The natural frequency of the swing is f_0 and the hinges supporting the ropes of the swing are rusty.

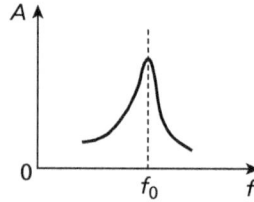

14 Which of the following graphs BEST illustrates the relation between the amplitude and driving frequency when the hinges are oiled?

(A)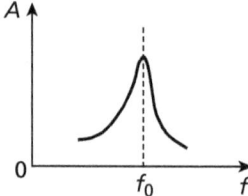

(C)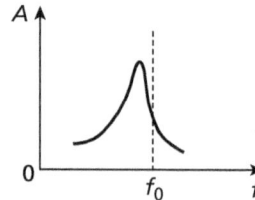

(B)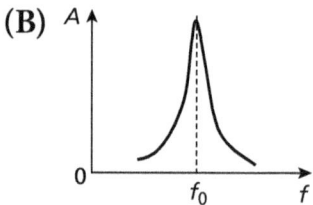

(D)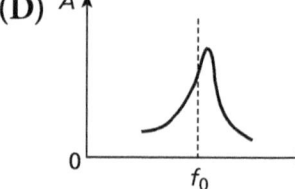

Ⓐ Ⓑ Ⓒ Ⓓ

15 Which of the following is/are true of a mass oscillating with simple harmonic motion?

   I. The acceleration is directly proportional to its displacement from its mean position.

   II. The velocity is proportional to its displacement from its mean position.

   III. The amplitude of the oscillation does not affect its frequency.

(A) I only

(B) I and II only

(C) I and III only

(D) III only

16 A mass hanging at rest from the lower end of a vertical spring is given a small vertical displacement and is then released at time $t = 0$. Which graph illustrates how the kinetic energy E_k and total energy E_T of the system vary with time t if it oscillates with simple harmonic motion?

(A)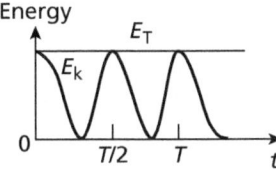

(C)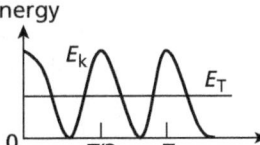

(B)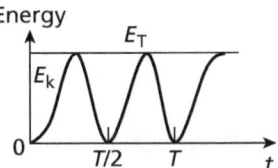

(D)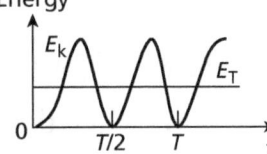

17 Which of the following is true for a body freely oscillating with simple harmonic motion?

(A) The angular frequency, the amplitude and the acceleration are all constant.

(B) If the amplitude of the vibration is increased, the period increases.

(C) The velocity is proportional to the displacement from the centre of the oscillation.

(D) The amplitude of the vibration does not affect its frequency.

1.1.5: Harmonic Motion (cont.)

18 Which of the following is NOT true of resonance?

(A) A large amplitude is obtained when the applied frequency is close to the natural frequency.

(B) It occurs when the driving frequency is greater than a certain critical frequency.

(C) The amplitude of a resonating body can be reduced by damping.

(D) When a system resonates at its maximum amplitude the rate of energy input to the system is equal to the rate of energy lost by the system.

19 The diagram below shows a system of pendulums. When the larger bob X is set oscillating the other bobs soon begin their own oscillations. Which small bob will obtain the greatest amplitude of vibration?

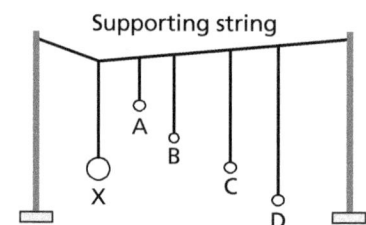

20 Which of the following is NOT true of damping with respect to oscillating systems?

(A) Critical damping occurs when the oscillation ceases in the shortest possible time.

(B) Moving coil ammeters and voltmeters, suspension systems and the foundations of structures all utilise the effects of damping.

(C) A vibrating guitar string is only slightly damped.

(D) Damping is useful since it increases the amplitude of a resonating body.

21 The graphs show how the amplitude A diminishes with time t as a body oscillates with varying degrees of damping. If the graphs have similar scales, which graph BEST indicates critical damping?

(A)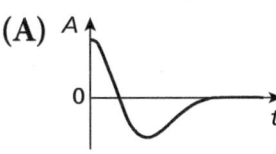

(C)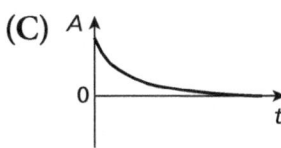

(B)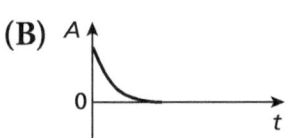

(D)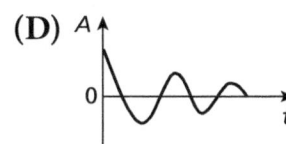

Items **22–23** refer to the following situation.

A particle oscillates with simple harmonic motion of amplitude 4.0 cm and angular frequency 10π. The timing of its oscillations is started at its maximum positive displacement.

22 What is its displacement at time $t = 0.015$ s?

(A) 1.8 cm

(B) −1.8 cm

(C) 3.6 cm

(D) −3.6 cm

23 What is its velocity at $t = 0.015$ s?

(A) 1.1 m s^{-1}

(B) -1.1 m s^{-1}

(C) -0.57 m s^{-1}

(D) 0.57 m s^{-1}

1.2.2: Properties of Waves 1

Item 1 refers to the following graph, which shows how the displacement x varies with time t for two waves.

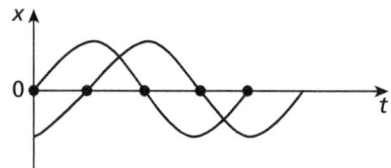

1 What is the phase difference, measured in radians, between the vibrations shown?

(A) $\frac{\pi}{4}$

(B) $\frac{\pi}{2}$

(C) $\frac{3\pi}{4}$

(D) π

2 A progressive wave of frequency 2.0 Hz is travelling at 16.0 m s^{-1}. What is the phase difference between points which are 3.0 m apart along the direction of progression of the wave?

(A) $\frac{3\pi}{4}$

(B) $\frac{3\pi}{8}$

(C) $\frac{3\pi}{2}$

(D) 3π

Item 3 refers to the following diagram, which shows a cross-section of a water wave travelling along the direction indicated.

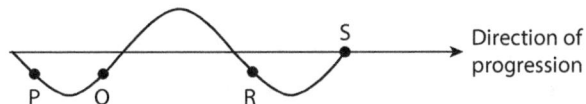

3. Which of the following describes the motion of P, Q, R and S at the instant shown?

	P	Q	R	S
(A)	down	up	down	stationary
(B)	up	down	up	down
(C)	left	stationary	left	right
(D)	up	down	up	stationary

4. Which of the following is/are true?

   I. Frequency determines the pitch of a sound wave and the colour of a light wave.

   II. Amplitude determines the loudness of a sound wave and the brightness of a light wave.

   III. High pitch notes travel faster than low pitch notes through the same medium.

   (A) I only
   (B) I and II only
   (C) II only
   (D) I, II and III

5. Which of the following is NOT true for a stationary wave?

   (A) Alternate nodes and antinodes exist along the line of its progression.
   (B) The distance between two successive nodes is equal to one half of a wavelength and the oscillations of adjacent particles between these nodes are in phase.
   (C) The oscillations of particles within adjacent segments between successive nodes are in antiphase.
   (D) It transmits energy along its direction of progression.

1.2.2: Properties of Waves 1 (cont.)

6 A standing wave containing 8 segments is set up in a string of length 12 m. What is the wave velocity in the string if the frequency is 10 Hz?

(A) 80 m s^{-1}

(B) 15 m s^{-1}

(C) 30 m s^{-1}

(D) 40 m s^{-1}

7 A speaker emits sound waves perpendicularly onto a hard reflecting surface and a stationary wave is set up between X and Y. A microphone connected to an oscilloscope is used to detect several nodes when moved between X and Y. It is moved from an initial node through 8 more nodes. If the speed of sound is 360 m s^{-1} and the microphone was moved 1.2 m, what is the frequency of the sound wave?

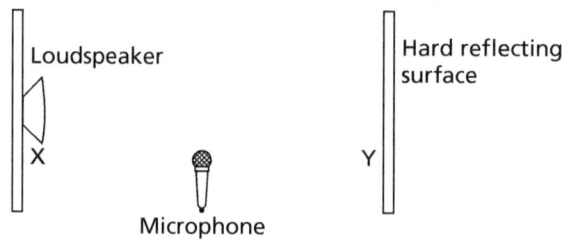

(A) 1200 Hz

(B) 600 Hz

(C) 300 Hz

(D) 38 Hz

8 A source of sound is placed at the open end of a glass tube which is closed at the other end. If sound travels at 340 m s^{-1} and ignoring any end correction, what must be the length of the tube for the fundamental resonant frequency to be 170 Hz?

(A) 0.25 m

(B) 1.0 m

(C) 2.0 m

(D) 0.50 m

9 The wavelengths of three sound waves X, Y and Z travelling in the same medium are 4 m, $\frac{4}{3}$ m and 2 m respectively. The ratio of their frequencies $f_X:f_Y:f_Z$ is therefore

(A) 1:2:3

(B) 3:2:1

(C) 1:3:2

(D) 2:1:3

10 Two sound waves approach each other to produce a stationary wave. Which of the following is/are true?

    I. They must be travelling in opposite directions along the same line.

    II. The distance L between their sources is given by $L = N\frac{\lambda}{2}$ where λ = wavelength and N is a positive integer.

    III. They must have the same waveform, wavelength and amplitude.

(A) I and II only

(B) I and III only

(C) II and III only

(D) I, II and III

Item **11** refers to the following diagram, which shows a stationary wave on a string at an instant in time.

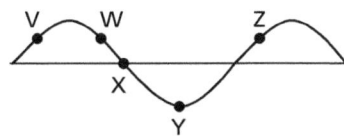

11 Which of the following is NOT true of the particles V, W, X, Y and Z?

(A) They vibrate with the same amplitude.

(B) W and Z vibrate in phase.

(C) V and W have the same speed and direction.

(D) Y is at an antinode and the acceleration of X is zero.

1.2.2: Properties of Waves 1 (cont.)

Item 12 refers to the following.

A loudspeaker connected to a signal generator is placed at the upper open end of a tall vertical glass tube which contains a small volume of water. The frequency f of the generator is increased from zero until the first instance of resonance occurs. The distance D from the water surface to the top of the tube is measured and recorded. By adding more water to the tube, several values of D and corresponding fundamental frequency f are found. The end correction and speed of sound are respectively e and v. A graph is plotted as shown below.

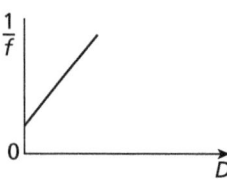

12 The gradient of the graph is

(A) $\dfrac{2}{v}$

(B) $\dfrac{4}{v}$

(C) $\dfrac{4e}{v}$

(D) $\dfrac{e}{v}$

Item 13 refers to the following diagram. Stationary waves are set up in a tube closed at one end. W, X, Y and Z are locations along the tube showing successive displacement nodes and antinodes.

13 Which of the following is/are true?

    I. Z is a displacement antinode and W is a pressure antinode.

    II. The particles at W, X, Y and Z all move back and forth in a direction parallel to the length of the tube.

    III. The particle at Y is stationary.

(A) I only

(B) I and II only

(C) II and III only

(D) I and III only

14 The amplitude of a wave at a distance of 5.0 m from a small point source is 0.20 mm. What would be the amplitude if the distance from the source is increased to 20 m?

(A) 0.80 mm

(B) 0.050 mm

(C) 3.2 mm

(D) 0.013 mm

15 The intensity of a wave at a point is 5.0 W m^{-2} when the amplitude of the wave there is 2.0 mm. What is the amplitude at the point when the intensity is increased to 22.5 W m^{-2}?

(A) 3.0 mm

(B) 9.0 mm

(C) 4.2 mm

(D) 18 mm

16 A small point source of power 5.0 W emits waves through a uniform and non-absorbing medium. What is the intensity received at a point 2.0 m from the source?

(A) 10 W m^{-2}

(B) 0.20 W m^{-2}

(C) 0.31 W m^{-2}

(D) 0.10 W m^{-2}

17 A small point source emits electromagnetic waves through a vacuum. At a distance of 5.0 m from the source the intensity received is 20 W m^{-2}. What is the intensity 3.0 m from the source?

(A) 12 W m^{-2}

(B) 7.2 W m^{-2}

(C) 33 W m^{-2}

(D) 56 W m^{-2}

1.2.2: Properties of Waves 1 (cont.)

18 Which of the following statements is/are true with respect to waves?

  I. Intensity at a point is proportional to the square of the amplitude at that point.

  II. Intensity received at a point from a point source emitting through a non-absorbing medium is inversely proportional to the square of the distance from the source.

  III. The amplitude received at a point from a point source emitting through a non-absorbing medium is inversely proportional to the distance from the source.

(A) I and II only

(B) I and III only

(C) II and III only

(D) I, II and III

Item 19 refers to the two graphs below, which show how the displacement x of a particle in a wave varies with position s relative to the source and with time t.

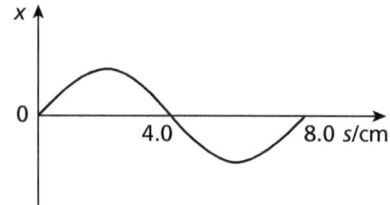 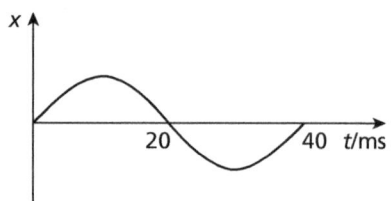

19 What is the speed of the wave?

(A) 2.0 m s^{-1}

(B) 0.20 m s^{-1}

(C) 240 m s^{-1}

(D) $3.2 \times 10^{-3} \text{ m s}^{-1}$

20 Red light of wavelength 7.0×10^{-7} m is emitted in pulses of 21 ms from an emergency system on a ship. What is the number of wavelengths in each pulse?

(A) 2.0×10^{18}

(B) 9.0×10^{15}

(C) 2.0×10^{16}

(D) 9.0×10^{12}

21 Which of the following statements about waves is/are true?

  I. All waves can exhibit reflection, refraction, diffraction and interference but only transverse waves can exhibit polarisation.

  II. The wavelength of a radio wave, a light wave and a gamma wave can be respectively 4.0×10^3 m, 4.0×10^{-7} m and 4.0×10^{20} m.

  III. X-rays are only slightly deflected by magnetic fields.

(A) I only

(B) I and II only

(C) I and III only

(D) II only

22 Light travels from material P of refractive index 2.0 to material Q of refractive index 1.6. The speed of light in P is 1.5×10^8 m s^{-1}. What is the speed of light in Q?

(A) 1.9×10^8 m s^{-1}

(B) 4.8×10^7 m s^{-1}

(C) 1.9×10^7 m s^{-1}

(D) 4.8×10^8 m s^{-1}

23 Light of wavelength 5.0×10^{-7} m s^{-1} travels from air to material Q, where its speed reduces to 1.8×10^8 m s^{-1}. The wavelength in Q is

(A) 8.3×10^{-7} m

(B) 3.0×10^{-7} m

(C) 1.2×10^6 m

(D) 3.0×10^7 m

1.2.2: Properties of Waves 1 (cont.)

24 Observer A stands 1050 m in front of a cliff and fires a shot. Observer B stands 350 m in front of the same cliff and hears two blasts including the returning echo. If the speed of sound is 350 m s^{-1} what is the time interval between the blasts heard by observer B?

(A) 4.0 s
(B) 3.0 s
(C) 2.0 s
(D) 1.0 s

25 Which of the following MUST be true when a wave refracts into a second medium and its speed decreases?

   I. The wavelength also decreases.
   II. The medium must be of greater density.
   III. The frequency increases.

(A) I only
(B) I and II only
(C) II and III only
(D) I and III only

26 Which of the following statements is/are true with respect to critical angle and total internal reflection of waves?

   I. Total internal reflection can only occur within the medium of greater density.
   II. For an interface between mediums A and B, total internal reflection can occur in A only if the wave can travel faster in B than it can in A.
   III. Total internal reflection can occur only if the angle of approach to the interface between the two mediums is greater than the critical angle of the medium it is in relative to the next medium.

(A) I and II only
(B) II and III only
(C) I, II and III
(D) III only

27 For light waves, mediums X and Y are of refractive index 1.5 and 1.4 respectively. What is the critical angle of X with respect to Y for light waves?

(A) 69°

(B) 42°

(C) 49°

(D) 43°

28 The wavelength of blue light in air is 4.0×10^{-7} m s^{-1}. What is the wavelength in X if the critical angle of X is 41.8°?

(A) 6.0×10^{-7} m

(B) 2.7×10^{-7} m

(C) 1.7×10^{-5} m

(D) 9.6×10^{-9} m

29 Which of the following is NOT true about optical fibres?

(A) Hundreds of thousands of telephone calls and hundreds of TV channels can be transmitted simultaneously through a single optical fibre.

(B) Digital signals through optical fibres are not as easily degraded as are analogue signals through copper wire.

(C) Endoscopes use bundles of optical fibres which transmit light into and out of places in the body which are otherwise difficult to reach.

(D) Optical fibres used in communication systems are usually comprised of a core and a cladding, the latter being of greater refractive index.

1.2.3: Properties of Waves 2

1 P and Q are point sources emitting waves of speed 200 mm s^{-1} and frequency 5.0 Hz. What are the phase differences between these waves at a point 240 mm from P and 260 mm from Q if they are emitted from the sources in phase and then in antiphase?

	in phase	in antiphase
(A)	0 radians	$\frac{\pi}{2}$ radians
(B)	$\frac{\pi}{4}$ radians	0 radians
(C)	π radians	0 radians
(D)	0 radians	π radians

2 P and Q are point sources emitting waves of wavelength 6.0 cm and of the same frequency. When emitting individually, they each produce vibrations of amplitude of 1.0 cm at point X. What is the amplitude of the vibrations at X when both sources are emitting waves and the phase difference between their emissions is π radians?

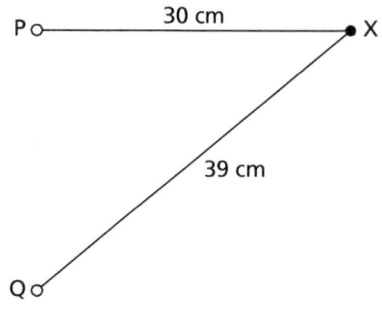

(A) 0 cm

(B) 1.0 cm

(C) 1.5 cm

(D) 2.0 cm

3 P and Q are point sources emitting waves of the same frequency but with a phase difference of 180°. The waves arrive at a point X with amplitudes of 7 cm and 2 cm respectively. If the distance between each source and X is the same, which of the following displacement–time graphs shows the vibration produced at that point?

(A)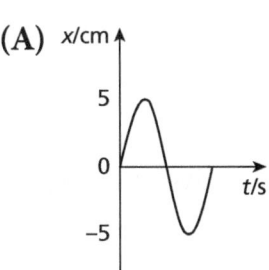

(B)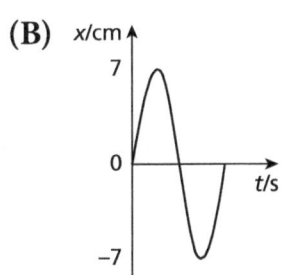

(C)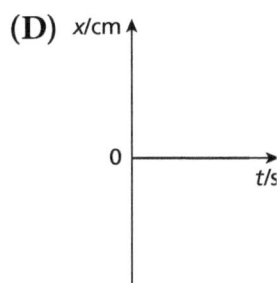

(D)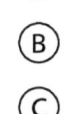

4 Which of the following statements about waves is/are true?

  I. According to the principle of superposition of waves, when waves meet at a point, their resultant displacement is the sum of their individual displacements.

  II. For sources to be coherent they must be in phase and emit waves of the same frequency.

  III. It is possible for coherent sources to be emitting waves with a phase difference of 180°.

(A) I and II only

(B) II only

(C) I and III only

(D) III only

1.2.3: Properties of Waves 2 (cont.)

5 Using a Young's double slit apparatus, which of the following statements are true for a dark fringe to be produced on the screen?

　　I. The waves meeting at the fringe must have a constant phase difference of π radians.

　　II. The sources must be coherent and must emit waves which arrive at the screen with the same amplitude.

　　III. The waves meeting at the dark fringe must have a path difference equal to a whole number of wavelengths.

(A) I and II only

(B) I and III only

(C) II and III only

(D) I, II and III

6 Two waves, A and B, of the same frequency and wavelength, have intensities of 25 W m^{-2} and 9.0 W m^{-2} respectively. What is the ratio of the amplitude of A to that of B?

(A) $\frac{25}{9}$

(B) $\frac{9}{25}$

(C) $\frac{3}{5}$

(D) $\frac{5}{3}$

7 An interference pattern was set up using yellow light with a Young's double slit apparatus. An adjustment was made which decreased the separation of the observed fringes. Which of the following adjustments could have caused this?

(A) Increasing the distance between the slits and screen.

(B) Using blue light instead of yellow light.

(C) Using a brighter source.

(D) Decreasing the distance between the slits.

8 A Young's double slit apparatus using light of wavelength 4.0×10^{-7} m produces a fringe separation of 3.0 mm. The distance between the slits and screen is halved, the slit separation is doubled and the source is replaced by one of wavelength 7.0×10^{-7} m. What is the new fringe separation?

(A) 1.7 mm

(B) 1.3 mm

(C) 5.3 mm

(D) 0.33 mm

9 Which of the following is NOT true?

(A) Increasing the number of slits in a diffraction grating decreases the width and increases the brightness of the fringes.

(B) The colours observed on a soap bubble exposed to sunlight are due to interference.

(C) Using a diffraction grating instead of just two slits has the advantage of producing more accurate measurement of the separation between fringes.

(D) When using a source of white light on a diffraction grating the nearest bright fringe to the central fringe is red.

10 Light of wavelength 450 nm is incident at right angles to a grating producing second order images with an angle of 70° between them. What is the number of lines per cm ruled on the grating?

(A) 6.4×10^{3} cm^{-1}

(B) 6.4×10^{5} cm^{-1}

(C) 1.3×10^{6} m^{-1}

(D) 1.3×10^{3} m^{-1}

1.2.3: Properties of Waves 2 (cont.)

11 What is the highest order image which can be seen on the screen of a diffraction apparatus if light of wavelength 500 nm is used and the grating has 6.4×10^5 rulings per metre?

(A) 6

(B) 5

(C) 4

(D) 3

Items **12–13** refer to the following situation.

Light waves X and Y, of wavelengths 4.0×10^{-7} m and 6.0×10^{-7} m respectively, are incident normally on a diffraction grating that has a ruling of 500 lines per mm.

12 In what order do the fringes overlap?

(A) 3rd order of Y overlaps 2nd order of X.

(B) 5th order of Y overlaps 3rd order of X.

(C) 3rd order of Y overlaps 5th order of X.

(D) 2nd order of Y overlaps 3rd order of X.

13 At what angle with the normal does the 3rd order of X occur?

(A) 64°

(B) 37°

(C) 24°

(D) 45°

14 The angle between the central line and the third order diffraction line seen on a screen is 40° when light of wavelength 500 nm is incident normally on a diffraction grating. What is the number of lines per mm ruled on the grating?

(A) 1.3×10^2 mm^{-1}

(B) 4.3×10^5 mm^{-1}

(C) 4.3×10^2 mm^{-1}

(D) 1.3×10^5 mm^{-1}

15 A Young's double slit experiment is set up using light of wavelength 500 nm and a distance of 1.2 m between the slits and the screen. What is the slit separation if the 1st bright fringe and the 3rd dark fringe from the central bright fringe are separated by a distance of 6.0 mm?

(A) 0.15 mm

(B) 0.40 mm

(C) 1.5×10^{-7} m

(D) 4.0×10^{-7} m

1.2.4: Physics of the Ear and Eye

1 Which of the following is/are true in relation to sound?

  I. Sound intensity level is a logarithmic measure of sound intensity in comparison to a reference level.

  II. For an average human the thresholds of hearing, feeling and pain are respectively 1×10^{-12} W m^{-2}, 1×10^0 W m^{-2} and 1×10^2 W m^{-2}.

  III. Loudness is the subjective response of the ear to an intensity level.

(A) I and II only

(B) I and III only

(C) II and III only

(D) I, II and III

1.2.4: Physics of the Ear and Eye (cont.)

2 A certain person can only detect sounds of frequencies between 50.0 Hz and 14 000 Hz. Which of the following pairs of wavelengths of sound can they detect when the speed of sound is 350 m s^{-1}?

(A) between 0.010 m and 6.9 m

(B) between 0.030 m and 4.2 m

(C) between 0.060 m and 8.4 m

(D) between 0.020 m and 7.2 m

3 What is the difference in intensity levels of two motors which emit sounds of intensities 1×10^{-5} W m^{-2} and 1×10^{-2} W m^{-2}?

(A) 3.0 dB

(B) 2.2 dB

(C) 30 dB

(D) 22 dB

4 What is the intensity of sound when the intensity level received from an explosion is 120 dB?

(A) 1.0 W m^{-2}

(B) 21 W m^{-2}

(C) 10 W m^{-2}

(D) 12 W m^{-2}

5 The ratio of the intensity received at a point for two sounds is 5:3. What is their difference in intensity levels?

(A) 0.22 dB

(B) 2.2 dB

(C) 17 dB

(D) 1.7 dB

6 Which of the following is/are true in relation to sound?

  I. If intensity is doubled, the intensity level is also doubled.

  II. Sensitivity is the ability of the ear to detect the smallest fractional change ΔI in an intensity I.

  III. Frequency response is the smallest change in frequency distinguishable by the ear at a given frequency.

(A) I and II only

(B) II only

(C) II and III only

(D) I, II and III

7 Which of the following is/are true in relation to noise?

  I. Noise is sound that is a nuisance to the person perceiving it.

  II. One may find that a mosquito creates more noise than a rock band.

  III. Noise is a very loud sound.

(A) I only

(B) I and II only

(C) I and III only

(D) I, II and III

8 Which of the following is/are true with respect to the eye?

  I. Accommodation is greatest when the eye views a distant object.

  II. Accommodation is the ability of the eye to alter the power of its lens.

  III. Depth of focus is the distance over which the image can be formed and still be considered to be in focus on the retina.

(A) I and II only

(B) II only

(C) I and III only

(D) II and III only

1.2.4: Physics of the Ear and Eye (cont.)

9 Which of the following is/are true in relation to problems associated with the eye?

  I. Astigmatism develops with age when the cells near the centre of the lens become under-nourished, die and become white.

  II. A person suffering from cataract has difficulty focusing light entering the eye in different planes.

  III. Myopia or short sight is a condition in which a person cannot see distant objects clearly.

(A) I only

(B) I and II only

(C) II and III only

(D) III only

10 A man cannot see objects clearly if they are closer than 40 cm from his eye. What is the power of the spectacle lens which can allow him to see objects at the normal near point of 25 cm?

(A) 0.67 D

(B) 1.5 D

(C) 0.015 D

(D) 67 D

11 A student cannot see objects clearly if they are more than 1.5 m from his eye. What is the focal length of the concave spectacle lens which can allow him to see distant objects clearly?

(A) 1.5 m

(B) 0.67 m

(C) 0.50 m

(D) 0.30 m

12 An object of height 5.0 cm is placed 50 cm in front of a converging lens of focal length 20 cm. Which of the following is/are true of the image produced?

    I. It is real and inverted and is 33.3 cm from the lens.

   II. Its magnification is 0.67.

  III. It is on the same side of the lens as is the object.

(A) I only

(B) I and II only

(C) I and III only

(D) II and III only

13 A point object is 15 cm in front of a concave lens of focal length 60 cm. Which of the following is true of the image?

(A) It is inverted and 5 cm from the lens on the side of the lens opposite to the object.

(B) It is not inverted and 12 cm from the lens on the same side as is the object.

(C) It is not inverted and 5 cm from the lens on the same side as is the object.

(D) It is inverted and 12 cm from the lens on the same side as is the object.

14 Which of the following is NOT true of the image formed by a converging lens of focal length F?

(A) It is real, diminished and inverted if the object is further from the lens than $2F$.

(B) It is real, diminished and inverted if the object is between F and $2F$ from the lens.

(C) It is virtual, magnified and erect if the object is closer to the lens than F.

(D) It is the same size as the object if placed at a distance $2F$ from the lens.

Module 3: Thermal and Mechanical Properties of Matter

1.3.1: Design and Use of Thermometers

1. A resistance thermometer registers a resistance of 28.2 Ω at the ice point and 75.6 Ω at the steam point. When placed in hot liquid the reading on the thermometer is 70.2 Ω. On the centigrade scale of this thermometer, the temperature of the liquid is

(A) 93 °C

(B) 88.6 °C

(C) 113 °C

(D) 47.4 °C

Ⓐ Ⓑ Ⓒ Ⓓ

2. A temperature on the absolute or thermodynamic scale is 525.23 K.
This is equivalent to

(A) 252.08 °C

(B) 252.07 °C

(C) 798.38 °C

(D) 798.39 °C

Ⓐ Ⓑ Ⓒ Ⓓ

3. The length of the thread in the bore of a large mercury thermometer is 25.2 cm at the ice point and 87.5 cm at the steam point. When its bulb is placed in a hot gel the reading is 75.0 °C. What is the length of the thread at this temperature?

(A) 80.0 cm

(B) 46.7 cm

(C) 85.7 cm

(D) 71.9 cm

Ⓐ Ⓑ Ⓒ Ⓓ

4 Which is NOT true of the following thermometers?

  I. A thermocouple responds quickly to temperature change, is of average accuracy, and can measure temperatures approximately between 30 K and 1750 K

  II. Both the constant volume gas thermometer and the resistance thermometer respond slowly to temperature change but are accurate, the former being very accurate. They can measure temperatures approximately between 3 K and 1750 K and between 75 K and 1550 K respectively.

  III. A mercury-in-glass thermometer responds rapidly to temperature change.

(A) I only

(B) I and II only

(C) II only

(D) III only

5 Which of the following thermometers have respectively the smallest and greatest temperature range?

(A) Mercury-in-glass, resistance

(B) Thermocouple, platinum resistance

(C) Mercury-in-glass, constant volume

(D) Alcohol-in-glass, mercury-in-glass

6 The following graph of resistance R against temperature T refers to a resistance thermometer. What is the temperature when the resistance is 53.4 Ω?

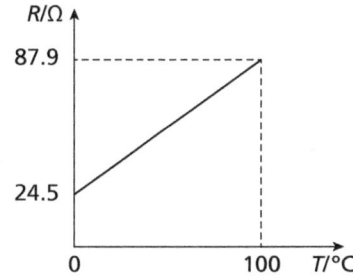

(A) 27.9 °C

(B) 63.4 °C

(C) 45.6 °C

(D) 60.1 °C

1.3.1: Design and Use of Thermometers (cont.)

7 Which of the following graphs shows how temperature θ_X on the centigrade scale varies with pressure p_T for a constant volume gas thermometer?

(A)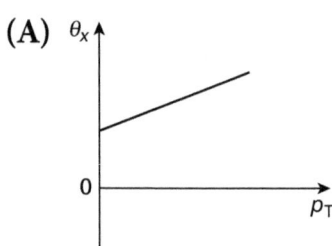

(C)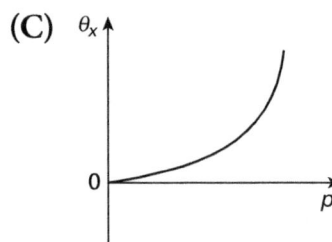

(B)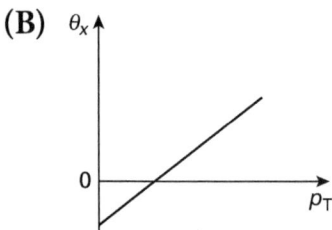

(D)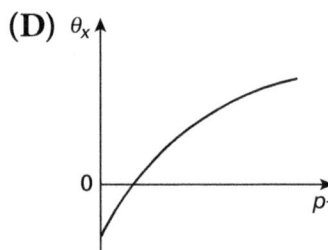

8 The reading on a resistance thermometer at the triple point of water is 29.42 Ω. What is the temperature on the thermodynamic scale when it is placed in a liquid and produces a reading of 43.52 Ω?

(A) 404.1 K

(B) 184.7 K

(C) 19.37 K

(D) 131.1 K

9 The thermometer BEST suited to measuring the slow variation of temperature several metres below the surface of the Earth is a

(A) constant volume gas thermometer.

(B) liquid-in-glass mercury thermometer.

(C) resistance thermometer.

(D) constant pressure gas thermometer.

1.3.2: Thermal Properties

Items **1–2** refer to the following graph, which shows how the temperature θ of a substance, initially solid and of mass 500 g, varies with time t as it receives energy from an electric heater.

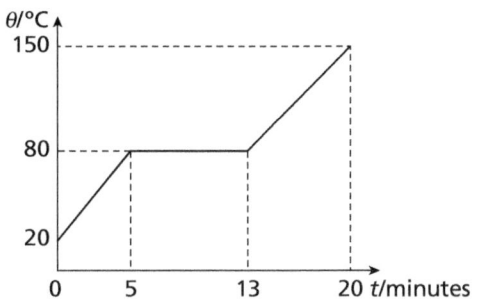

1 The specific heat capacity of the solid is 2000 J kg^{-1} K^{-1}. What is the power of the heater?

(A) 200 W

(B) 12 kW

(C) 24 kW

(D) 400 W

2 What is the specific latent heat of fusion of the substance?

(A) 3.84×10^5 J kg^{-1}

(B) 1.92×10^5 J kg^{-1}

(C) 5.8×10^6 J kg^{-1}

(D) 5.8×10^6 J g^{-1}

3 At a certain location thermal radiation reaches the Earth at a power of 800 W m^{-2}. It is incident on a collecting heater panel of area 8.0 m^2 of a solar water heater which is only capable of transferring 20% of the received energy to the water it warms. How long will it take to raise the temperature of 30 kg of the water by 50 °C?

(A) 16 minutes

(B) 82 minutes

(C) 410 minutes

(D) 656 minutes

1.3.2: Thermal Properties (cont.)

4 A water drop at 5.0 °C and of mass 4.0 mg falls onto a large frozen pond. What energy is released by the water drop if the pond is at 0 °C?

(A) 1.42 kJ

(B) 1.4 J

(C) 0.084 kJ

(D) 1.34 J

5 A power supply of 120 V is applied across a resistor of heat capacity 200 J K^{-1} and of mass 250 g. What is the rise in temperature in 30 s if a steady current of 4.0 A flows?

(A) 18 °C

(B) 72 °C

(C) 288 °C

(D) 36 °C

6 Which of the following statements is/are true?

I. The heat capacity of a body is the heat needed to raise the temperature of 1 kg of the body by 1 °C.

II. The latent heat of fusion of a body is the heat needed to change 1 kg of the body from solid to liquid without a change of temperature.

III. It is possible for heat to flow from a body X to a body Y even if body X contains less thermal energy than body Y.

(A) I and II only

(B) II only

(C) I, II and III

(D) III only

7 The graph below shows the variation of temperature θ with time t for a substance being cooled at a constant rate.

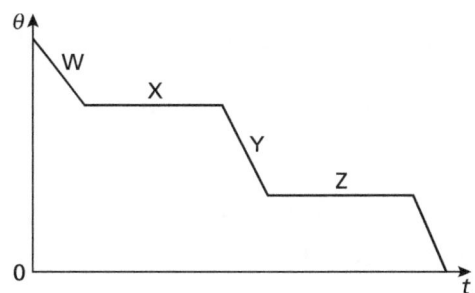

Which of the following BEST describes what occurs during the stages W, X, Y and Z?

(A) W: The solid cools and the kinetic energy of its particles decreases. Ⓐ

(B) X: The substance changes from gas to liquid as its particles lose kinetic energy. Ⓑ

(C) Y: The liquid increases in temperature as its molecules obtain more kinetic energy. Ⓒ

(D) Z: The substance changes from liquid to solid as its particles lose potential energy. Ⓓ

8 The main reason that metals are better conductors of heat than non-metals is that

(A) metals are generally denser, making it easier to pass on the energy. Ⓐ

(B) metals contain many more electrons. Ⓑ

(C) metals form ionic compounds and are therefore composed of many electric charges. Ⓒ

(D) metal atoms have loosely bound electrons which can free themselves from a particular atom and transfer their kinetic energy to other atoms. Ⓓ

1.3.2: Thermal Properties (cont.)

9 Which of the following is/are true regarding energy transfer?

    I. Convection in fluids occurs due to differences in density existing within the fluid.

    II. If one is working in a large freezer room it is better to wear white than black in order to stay warm.

    III. Air pockets within the fibres of wool clothing primarily reduce the amount of radiation through the clothing.

(A) I only

(B) I and II only

(C) II and III only

(D) I, II and III

1.3.3: Heat Transfer

<u>Item 1</u> refers to a perfectly lagged bar comprised of materials X, Y and Z of conductivities k_X, k_Y and k_Z respectively. The graph shown indicates the variation of temperature θ with distance D from the hotter end of the bar when a temperature difference is set up across its length. Steady state conditions prevail and each material is of length 20 cm.

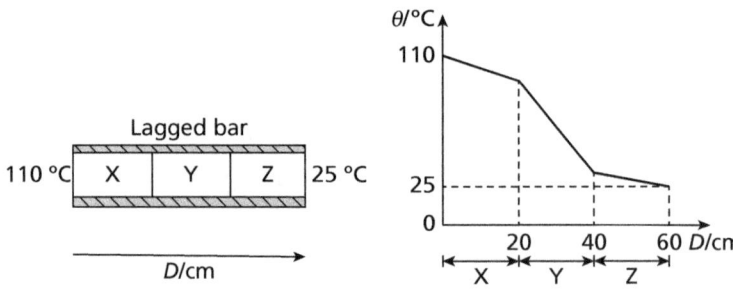

1 Which of the following is true of the conductivities of X, Y and Z?

(A) $k_X > k_Y > k_Z$

(B) $k_X > k_Z > k_Y$

(C) $k_Y > k_Z > k_X$

(D) $k_Z > k_X > k_Y$

2 The graph shows how temperature θ varies with the length x from the hotter end of a bar as heat conducts through it. Which of the following conditions could have resulted in this variation?

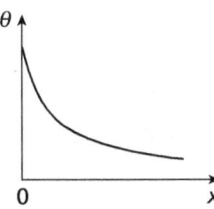

   I. The bar is perfectly lagged and steady state conditions prevail.
   II. The bar is not lagged and steady state conditions prevail.
   III. The bar is perfectly lagged but the heat has only just been applied to one of its ends and steady state conditions have not yet been attained.

(A) I only
(B) I and II only
(C) II and III only
(D) III only

Item 3 refers to the following diagram, which shows a perfectly lagged copper bar in which the cross-sectional area increases uniformly from X to Y.

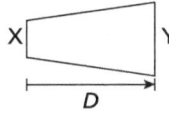

3 The bar is heated at X and steady state conditions are reached. Which of the following graphs BEST illustrates the variation of temperature θ with distance D along the length of the bar?

(A)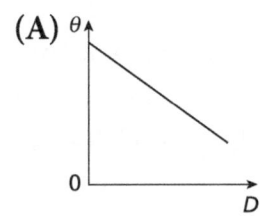

(C)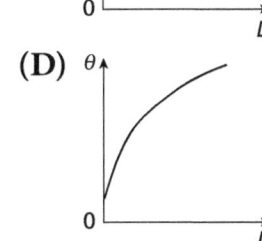

(B)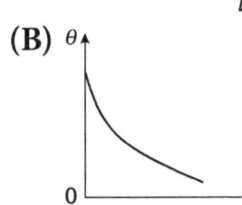

(D)

1.3.3: Heat Transfer (cont.)

Items **4–6** refer to the diagram below, which shows composite bars made from sections of two metals, P and Q, of conductivities of 400 W m^{-1} K^{-1} and 200 W m^{-1} K^{-1} respectively. P and Q are both of cross-sectional area 1.00×10^{-2} m^2. The structures are perfectly lagged and a temperature difference of 40 K is set up across their lengths.

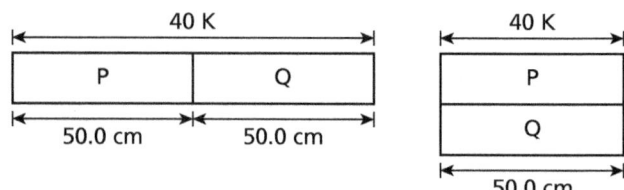

4 What is the rate of heat flow under steady state conditions with P and Q in series?

(A) 107 W

(B) 120 W

(C) 480 W

(D) 240 W

5 What is the rate of heat flow under steady state conditions with P and Q in parallel?

(A) 120 W

(B) 480 W

(C) 240 W

(D) 107 W

6 A single bar of material Q of the same cross-sectional area is to be used to allow the same rate of flow of heat through it as the composite series arrangement when the same temperature difference exists between its ends. What should be the length of the bar if this rate of flow is 107 W?

(A) 0.75 m

(B) 0.65 m

(C) 1.25 m

(D) 1.50 m

7 A small hot sphere at a temperature T_1 and of emissivity ε is placed in a large chamber with walls at a cooler temperature T_2. If the surface area of the sphere is A and the Stefan–Boltzmann constant is σ, what is the net thermal power initially emitted by the sphere?

(A) $P = \varepsilon \sigma A \left(T_2^4 - T_1^4 \right)$

(B) $P = \varepsilon A \left(T_1 - T_2 \right)^4$

(C) $P = \varepsilon A \left(T_2 - T_1 \right)^4$

(D) $P = \varepsilon \sigma A \left(T_1^4 - T_2^4 \right)$

8 A cube of side 1.00 cm, emissivity 0.50, and temperature 20 °C, is placed into an enclosure at a temperature of 180 °C. What is the net rate of absorption of thermal energy by the cube?

(A) 1.2 W

(B) 9.9×10^{-2} W

(C) 0.59 W

(D) 1.1×10^{-2} W

1.3.3: Heat Transfer (cont.)

Items **9–10** refer to the following diagram.

The metal bar is well lagged and has a uniform cross-sectional area of 8.0 cm^2. Energy is supplied to its left end by an electric heater of power 120 W and is removed from its right end by water circulating in a tube in contact with it. A steady state is maintained.

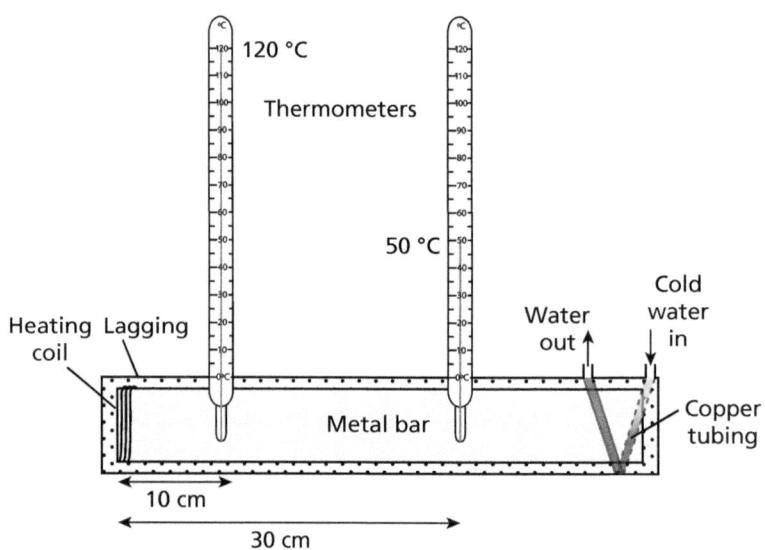

9 What is the value of the thermal conductivity of the metal?

(A) 4.3×10^2 W m^{-1} K^{-1}

(B) 4.3 W m^{-1} K^{-1}

(C) 4.3×10^6 W m^{-1} K^{-1}

(D) 4.3×10^4 W m^{-1} K^{-1}

10 The temperature of the water at the cool end rises by 4.0 K. What is the mass of water per minute flowing through the tubes if the specific heat capacity of water is 4.2 J g^{-1} K^{-1}?

(A) 114 g min^{-1}

(B) 7.1 g min^{-1}

(C) 430 g min^{-1}

(D) 126 g min^{-1}

11 A heater of power 500 W maintains a copper block at a steady temperature. The mass of the block is 2.50 kg and its specific heat capacity is 380 J kg^{-1} K^{-1}. What is the rate of fall in temperature when the heater is switched off?

(A) 3.29 K s^{-1}

(B) 76.0 K s^{-1}

(C) 0.53 K s^{-1}

(D) 3.04 K s^{-1}

12 Which of the following is/are true of a black body?

   I. A black body absorbs all radiation which is incident on it.

   II. A perfect absorber of radiation must also be a perfect emitter of radiation.

   III. The filament of a lamp in use is a black body emitter.

(A) I only

(B) II only

(C) I and II only

(D) II and III only

1.3.4: The Kinetic Theory of Gases

1 Which of the following is NOT an assumption of the kinetic theory of gases?

(A) Attraction between the particles is negligible since it only exists at the brief instant of collision. The potential energy is therefore zero.

(B) The particles occupy negligible space relative to the volume of their enclosure.

(C) The molecules of a given gas are not all identical.

(D) Collisions between the particles are elastic.

1.3.4: The Kinetic Theory of Gases (cont.)

2 Which of the following is NOT an assumption of the kinetic theory of gases?

(A) The impact time of collisions is small and negligible compared with the time spent travelling between collisions.

(B) The motion of the molecules is random.

(C) Kinetic energy is not conserved on collisions between particles.

(D) The number of molecules is large and therefore their average speed is statistically meaningful.

3 The mean kinetic energy of a single molecule of a monatomic gas at a temperature of 200 °C is

(A) 9.79×10^{-21} J

(B) 5.90×10^{3} J

(C) 6.53×10^{-21} J

(D) 4.14×10^{-21} J

4 The N molecules of an ideal gas travel at individual speeds $C_1, C_2, \ldots C_N$. Their r.m.s. speed is

(A) $\dfrac{\sqrt{C_1^2 + C_2^2 + \ldots C_N^2}}{N}$

(B) $\sqrt{\dfrac{C_1^2 + C_2^2 + \ldots C_N^2}{N}}$

(C) $\sqrt{\dfrac{(C_1 + C_2 + \ldots C_N)^2}{N}}$

(D) $\sqrt{\dfrac{C_1 + C_2 + \ldots C_N}{N}}$

5 The speeds of a group of 20 particles are as follows

3 particles at 5.0 m s^{-1}, 7 particles at 6.0 m s^{-1}, 8 particles at 7.0 m s^{-1}, 2 particles at 8.0 m s^{-1}. What is the r.m.s. speed of the particles?

(A) 6.5 m s^{-1}

(B) 29.1 m s^{-1}

(C) 189 m s^{-1}

(D) 2.54 m s^{-1}

6 Which of the following is NOT true of the molecules of a liquid?

(A) They are packed almost as close as in solids.

(B) The forces between molecules are extremely strong, causing them to have a fixed volume.

(C) The motion of their molecules is random.

(D) The intermolecular forces are not strong and therefore liquids do not have a fixed shape.

7 The graph shows how the pressure p of the air in a strong steel vessel varies with mass m as n moles of air are pumped into it. The temperature T and the volume V are constant.

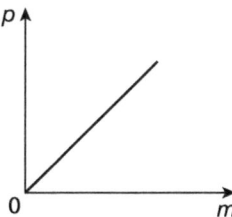

If the molar gas constant and molar mass are represented by R and M_0 respectively, then the gradient of the graph is

(A) $\dfrac{M_0 V}{RT}$

(B) $\dfrac{nRT}{V}$

(C) $\dfrac{RT}{M_0 V}$

(D) nRT

8 A gas of molar mass 4.0 g at a temperature of 27 °C has a pressure of 3.00×10^6 Pa and volume of 1.00×10^{-3} m^3. The number of moles of the gas and the number of molecules of the gas are respectively

(A) 13.3 7.2×10^{20}

(B) 1.2 28.8×10^{23}

(C) 1.2 7.2×10^{23}

(D) 13.3 28.8×10^{20}

1.3.4: The Kinetic Theory of Gases (cont.)

9 4.0 moles of argon of molar mass 39.9 g are kept in a vessel. The number of argon particles is

(A) 2.3×10^{26}

(B) 1.5×10^{23}

(C) 2.41×10^{24}

(D) 1.2×10^{20}

10 If a monatomic gas at a pressure of 2.00×10^5 Pa has an r.m.s. speed of 800 m s^{-1}, what is its density?

(A) 0.938 kg m^{-3}

(B) 750 kg m^{-3}

(C) 0.208 kg m^{-3}

(D) 7.5 kg m^{-3}

11 A further N moles of gas is added to a strong steel vessel initially containing 3.0 mol of gas at a pressure of 6.0×10^5 Pa and temperature 27 °C. After a few minutes the temperature adjusts to its initial value and the new pressure reached is 2.0×10^6 Pa. What is the value of N?

(A) 5

(B) 6

(C) 7

(D) 8

12 A vessel contains 1.5×10^{23} molecules of gas at a pressure of 4.0×10^5 Pa and a volume of 2.0×10^{-3} m^3. What is the temperature of the gas?

(A) 110 °C

(B) 386 °C

(C) 452 °C

(D) 772 °C

13 A monatomic gas at an absolute temperature T, pressure p and volume V contains N molecules comprising n moles of gas. Using the molar gas constant as R and the Boltzmann constant as k, which of the following pairs of equations can be used to determine the kinetic energy of a single molecule of the gas E_{k1} and the total kinetic energy E_k of the gas respectively?

(A) $E_{k1} = \frac{3}{2}kT$ $\quad\quad E_k = \frac{3}{2}NRT$

(B) $E_{k1} = \frac{3}{2}pV$ $\quad\quad E_k = \frac{3}{2}NkT$

(C) $E_{k1} = \frac{3}{2}nRT$ $\quad\quad E_k = \frac{3}{2}pV$

(D) $E_{k1} = \frac{3}{2}kT$ $\quad\quad E_k = \frac{3}{2}nRT$

14 Which of the following is/are true for an ideal gas?

   I. The number of molecules in 1 kg of a gas is 6.02×10^{23}.

   II. The molar gas constant is the product of the Boltzmann constant and Avogadro's number.

   III. One mole of argon has the same volume as one mole of helium at the same temperature and pressure.

(A) I only

(B) I and II only

(C) I and III only

(D) II and III only

15 What is the r.m.s speed of helium atoms at 100 K if the molar mass of helium is 4.0 g?

(A) 25 m s^{-1}

(B) 790 m s^{-1}

(C) 456 m s^{-1}

(D) 558 m s^{-1}

1.3.4: The Kinetic Theory of Gases (cont.)

16 The particles of a gas at 50.0 °C have an r.m.s. speed v. At what temperature will the r.m.s. speed be $3v$?

(A) 969 °C

(B) 2907 °C

(C) 2630 °C

(D) 696 °C

1.3.5: First Law of Thermodynamics

1 If the equation expressing the first law of thermodynamics is written as $\Delta U = \Delta Q + \Delta W$, where ΔU and ΔQ represent changes in internal energy and heat added to the system respectively, which of the following is true?

(A) ΔW represents work done by the gas.

(B) $\Delta U = 0$ if heat does not enter or leave the system.

(C) $\Delta Q = 0$ in an adiabatic process.

(D) $\Delta W = p\Delta V$ where p and ΔV represent the pressure and change in volume of the gas respectively.

2 Which of the following is NOT true of an ideal gas?

(A) Any changes in internal energy are due to a corresponding change in temperature.

(B) For an isothermal change, temperature and kinetic energy are constant.

(C) When a fixed mass of gas goes through a cyclic process, the change is path-independent and the net changes of pressure, volume and temperature are all zero.

(D) Positive work is done by the gas when it contracts.

Items **3–5** refer to the following *pV* diagram, which shows an ideal gas undergoing a cyclic process.

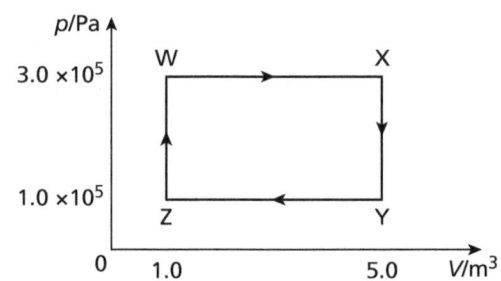

3. What is the work done BY the gas during XY and YZ respectively?

(A) 0 J -4.0×10^5 J
(B) 2.0×10^5 J 1.0×10^6 J
(C) 1.0×10^6 J -2.0×10^5 J
(D) 0 J 4.0×10^5 J

4. What is the net work done ON the gas for the complete cycle?

(A) 0 J
(B) -4.0×10^5 J
(C) -8.0×10^5 J
(D) 12×10^5 J

5. What are the values of ΔU and ΔQ for the cyclic process if the work done by the gas is 8.0×10^5 J?

(A) $\Delta U = -8.0 \times 10^5$ J $\Delta Q = 12 \times 10^5$ J
(B) $\Delta U = -8.0 \times 10^5$ J $\Delta Q = 0$
(C) $\Delta U = 0$ $\Delta Q = 8.0 \times 10^5$ J
(D) $\Delta U = 0$ $\Delta Q = 0$

1.3.5: First Law of Thermodynamics (cont.)

6 8.0 moles of gas is heated at a constant pressure of 2.0×10^5 Pa, from 300 K to 400 K. What is the heat supplied, given that the molar heat capacity at constant pressure is 29 J mol^{-1} K^{-1}?

(A) 2.3×10^4 J

(B) 4.6×10^9 J

(C) 4.6×10^5 J

(D) 2.3×10^9 J

7 1 kg of steam at 373 K and 1.0×10^5 Pa occupies a volume of 1.7 m^3 whereas 1 kg of water occupies an insignificant volume at the same temperature when compared to this. What is the work done by the gas on the formation of 50 g of steam at 100 °C and atmospheric pressure?

(A) 7.1×10^3 J

(B) 8.5×10^3 J

(C) 1.7×10^5 J

(D) 3.6×10^2 J

8 Which of the following is NOT true for n moles of a monatomic gas in relation to its molar heat capacities at constant pressure C_p and at constant volume C_V?

(A) $C_p = C_V + R$

(B) $C_p = \dfrac{3}{2} R$

(C) $\dfrac{3}{2} nR\Delta T = nC_V \Delta T$

(D) $C_V = \dfrac{3}{2} R$

9 Which of the following is/are NOT true of the gas undergoing the cyclic process shown?

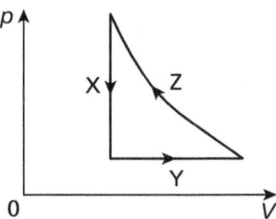

(A) The temperature falls, but no work is done during X.

(B) Positive work is done by the gas and the temperature rises during Y.

(C) During Z the gas is compressed and therefore its temperature must rise.

(D) Z may be an adiabatic change or an isothermal change.

10 Which of the following could describe the types of changes occurring in the cyclic process shown?

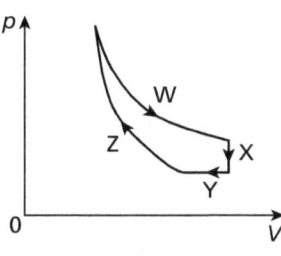

	W	X	Y	Z
(A)	adiabatic	isovolumetric	isobaric	isothermal
(B)	isothermal	isovolumetric	adiabatic	isobaric
(C)	isothermal	isobaric	isovolumetric	adiabatic
(D)	isothermal	isovolumetric	isobaric	adiabatic

1.3.5: First Law of Thermodynamics (cont.)

11 500 J of heat energy is added to a gas and it is allowed to expand by 0.002 m³ at a constant pressure of 1.2×10^5 Pa. The change in internal energy ΔU and the work done ΔW on the gas are

	ΔU	ΔW	
(A)	260 J	−240 J	Ⓐ
(B)	240 J	260 J	Ⓑ
(C)	−260 J	240 J	Ⓒ
(D)	−240 J	−260 J	Ⓓ

12 Which of the following is true during an adiabatic expansion?

(A) Work is done on the gas and heat is added to it. Ⓐ

(B) Work is done by the gas but no heat enters or leaves it. Ⓑ

(C) Work is done by the gas and heat leaves it. Ⓒ

(D) Work is done on the gas but no heat enters or leaves it. Ⓓ

13 The first law of thermodynamics states that $\Delta U = \Delta Q + \Delta W$, where ΔU is the change in internal energy, ΔQ the heat added, and ΔW the work done on the system. Which of the following must be **zero** for an isothermal change and for an adiabatic change?

	Isothermal change	Adiabatic change	
(A)	ΔU	ΔW	Ⓐ
(B)	ΔQ	ΔU	Ⓑ
(C)	ΔU	ΔQ	Ⓒ
(D)	ΔQ	ΔW	Ⓓ

14 The graph shows a gas undergoing a cyclic process.

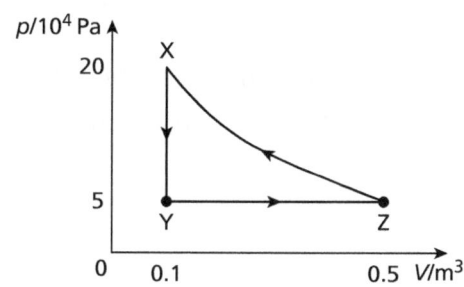

Which of the following is true of the temperatures T_X, T_Y and T_Z at X, Y and Z?

(A) $T_Z > T_X > T_Y$

(B) $T_X > T_Z > T_Y$

(C) $T_X > T_Y > T_Z$

(D) $T_Z > T_Y > T_X$

15 In a certain cyclic process, the heat supplied is Q_1 and the heat lost is Q_2. What is the efficiency of the cycle?

(A) $\dfrac{Q_2 - Q_1}{Q_1}$

(B) $\dfrac{Q_2}{Q_1}$

(C) $\dfrac{Q_1}{Q_2}$

(D) $\dfrac{Q_1 - Q_2}{Q_1}$

16 Air is pumped into an inflatable cube with elastic sides to be used as a float in a swimming pool. The atmospheric pressure on the day is 1.00×10^5 Pa. What is the work done against the atmospheric pressure as the length of the cube's side grows from 10.0 cm to 30.0 cm?

(A) 2.70×10^9 J

(B) 2.60×10^3 J

(C) 2.00×10^4 J

(D) 8.00×10^3 J

1.3.6: Mechanical Properties of Materials

1 The molar mass and density of aluminium are respectively 27.0 g and 2.70 g cm^{-3}. What is the volume required by an atom of aluminium in its lattice?

(A) 1.66×10^{-26} cm^{-3}

(B) 1.66×10^{-24} cm^{-3}

(C) 1.66×10^{-23} cm^{-3}

(D) 1.66×10^{-29} cm^{-3}

2 Metallic materials X and Y of densities 9000 kg m^{-3} and 5000 kg m^{-3} respectively are combined to make a new alloy. The mass of X used is twice that of Y. What is the density of the new alloy?

(A) 7105 kg m^{-3}

(B) 7000 kg m^{-3}

(C) 7530 kg m^{-3}

(D) 6900 kg m^{-3}

<u>Items 3–4</u> refer to the following diagram where a cube with side of length L floats in a liquid of density ρ with the liquid line half way up its height. When a smaller cube of mass m is placed on top of it, the liquid level changes such that only one quarter of the larger cube remains above the surface.

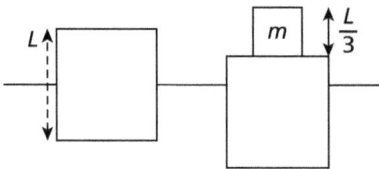

3 The mass of the smaller cube can be expressed as

(A) $4\rho L^3$

(B) $\dfrac{4\rho}{L^2}$

(C) $\dfrac{\rho L^3}{4}$

(D) $\dfrac{\rho L}{2}$

4 What is the density of the smaller cube?

(A) $\dfrac{27m}{L^3}$

(B) $27mL^3$

(C) $\dfrac{9m}{L^3}$

(D) $3mL^3$

5 A swimming pool of length 5.00 m and width 3.00 m has a uniform depth of 2.00 m. If the density of the water in the pool is 1050 kg m^{-3} and atmospheric pressure on the day is 1.04×10^5 Pa, what is the pressure on the floor of the pool?

(A) 2.1×10^4 Pa

(B) 1.25×10^5 Pa

(C) 4.13×10^5 Pa

(D) 2.1×10^3 Pa

6 The following diagram shows a conical flask containing water. What is the force F (**due to the water**) exerted on its base?

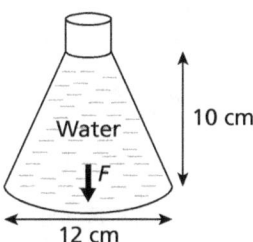

(A) 27 N

(B) 981 N

(C) 185 N

(D) 11 N

1.3.6: Mechanical Properties of Materials (cont.)

7 What is the pressure of the gas supply if the atmospheric pressure is 1.02×10^5 Pa and the liquid in the tube shown below is of density 7.5×10^3 kg m^{-3}?

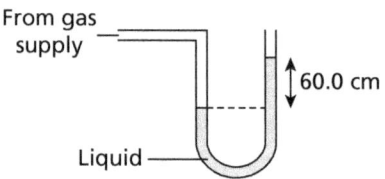

(A) 1.5×10^5 Pa

(B) 4.4×10^4 Pa

(C) 4.4×10^6 Pa

(D) 2.5×10^5 Pa

Items **8–10** refer to the following diagram. Some of the air is sucked out from the opened tap and then the tap is closed. Liquid Y is of density ρ_y and the acceleration due to gravity is g.

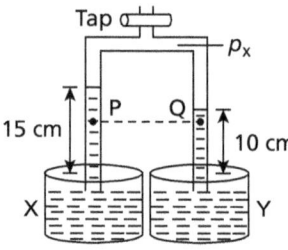

8 The density of liquid X is

(A) $\dfrac{2}{3}\rho_y$

(B) $\dfrac{2}{3}\rho_y g$

(C) $\dfrac{3}{2}\rho_y g$

(D) $\dfrac{3}{2}\rho_y$

9 If the atmospheric pressure is p_A, the pressure p_x above the liquids in the tubes is

(A) $p_A - 0.10\rho_y g$

(B) $p_A - 0.15\rho_y g$

(C) $p_A + 0.15\rho_y g$

(D) $p_A + 0.10\rho_y g$

10 Which of the following is/are true?

   I. The pressure at the top of each liquid column is the same.

   II. The pressure at the base of each liquid column is the same.

   III. The pressure at point P is equal to the pressure at point Q.

(A) I and II only

(B) II and III only

(C) I and III only

(D) I, II and III

11 The diameter of an air bubble on the bed of a lake rises and increases by a factor of 2 just before it breaks the surface where the atmospheric pressure is 1.0×10^5 Pa. What is the pressure at the bottom of the lake, assuming that the temperature of the bubble changes from 10 °C to 20 °C as it ascends?

(A) 2.0×10^5 Pa

(B) 1.9×10^5 Pa

(C) 3.9×10^5 Pa

(D) 7.7×10^5 Pa

12 The water in a lake is at uniform temperature and its density is 1.1×10^3 kg m^{-3}. The pressure on its bed is 5 times the value at the surface when the pressure of the atmosphere is 1.0×10^5 Pa. What is the depth of the lake?

(A) 41 m

(B) 46 m

(C) 74 m

(D) 37 m

1.3.6: Mechanical Properties of Materials (cont.)

Item 13 refers to the following diagram.

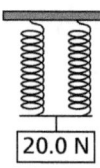

13. The springs are exactly the same and the extension produced is 5.0 cm. The force constants of a single spring and of the parallel combination are

	Single spring	Parallel springs
(A)	8.0 N cm^{-1}	4.0 N cm^{-1}
(B)	2.0 N cm^{-1}	4.0 N cm^{-1}
(C)	4.0 N cm^{-1}	2.0 N cm^{-1}
(D)	4.0 N cm^{-1}	8.0 N cm^{-1}

Items 14–15 refer to the following diagram.

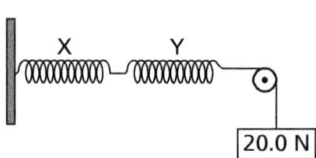

14. The springs are exactly the same and the total extension produced is 5.0 cm. The force constants of a single spring and of the series combination are:

	Single spring	Series combination
(A)	2.0 N cm^{-1}	4.0 N cm^{-1}
(B)	2.0 N cm^{-1}	8.0 N cm^{-1}
(C)	4.0 N cm^{-1}	2.0 N cm^{-1}
(D)	8.0 N cm^{-1}	4.0 N cm^{-1}

15 X and Y are replaced by P and Q. P has a force constant of 4.0 N cm^{-1} and it is found that the total extension is now 7.0 cm. The force constant of Q is therefore

(A) 10 N cm^{-1}

(B) 2.0 N cm^{-1}

(C) 5.0 N cm^{-1}

(D) 7.0 N cm^{-1}

Items **16–17** refer to the following graph, which shows the variation of force F with extension e as a material is loaded and then unloaded.

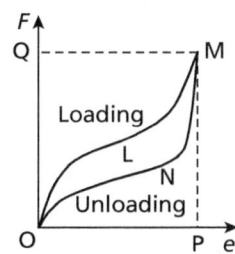

16 Which area represents the energy lost as a result of the cycle?

(A) OQMLO

(B) OQMNO

(C) OLMNO

(D) ONMPO

17 Which of the following could be the material under investigation?

(A) A glass fibre.

(B) A copper wire.

(C) A ceramic strip.

(D) A rubber strand.

1.3.6: Mechanical Properties of Materials (cont.)

18 A wire is subjected to a force F along its length L, which produces an extension e if the cross-sectional area of the wire is A. The Young modulus is

(A) $\dfrac{Fe}{LA}$

(B) $\dfrac{FL}{Ae}$

(C) $\dfrac{AL}{Fe}$

(D) $\dfrac{eA}{FL}$

19 The following graphs show the variation in extension as various loads are applied to materials P, Q and R. Which of the following describes, in the order P-Q-R, the type of each material and gives an example of it?

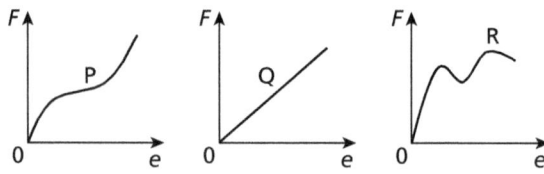

(A) ductile/copper   brittle/glass   polymeric/rubber

(B) polymeric/rubber   ductile/copper   brittle/glass

(C) brittle/copper   ductile/rubber   polymeric glass

(D) polymeric/rubber   brittle/glass   ductile/copper

20 Which of the following is/are true?

    I. Crystalline solids have a very regular arrangement of their atoms whereas amorphous solids do not.

    II. Natural rubber is a polymer with its molecules forming tangled strands. These strands straighten out when the rubber is stretched.

    III. Examples of crystalline solids are metals, diamond, silicon and other group IV elements.

    IV. Glass is not an amorphous solid.

(A) I and II only

(B) II and III only

(C) I, II and III only

(D) I, II and IV only

Items **21–23** refer to the following graph, which shows how the length L varies as a metal wire is loaded and then unloaded by a force F.

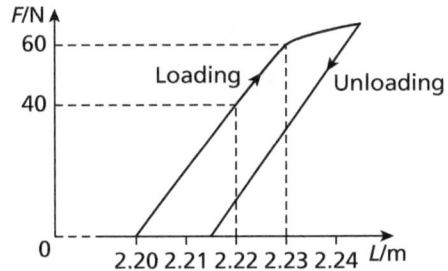

21 The potential energy stored when the load is 60 N is

(A) 9.0×10^{-3} J

(B) 0.54 J

(C) 0.90 J

(D) 0.54 J

22 Which of the following is NOT true?

(A) The wire has obtained a permanent stretch of about 1.5 cm.

(B) The proportional limit and the elastic limit have both been surpassed in the loading.

(C) The stress at the proportional limit is 60 N.

(D) The strain is 9.1×10^{-3} when a force of 40 N is applied during the loading.

23 How much work is done in stretching the wire when the force is increased from 40 N to 60 N on loading?

(A) 0.20 J

(B) 0.30 J

(C) 0.40 J

(D) 0.50 J

1.3.6: Mechanical Properties of Materials (cont.)

24 A ligament of length 30 cm stretches 9.0 mm when a force of 60.0 N is applied to it along its length. If the diameter of the ligament is 5.0 mm, its Young modulus is

(A) 1.0×10^8 Pa

(B) 1.0×10^5 Pa

(C) 1.0×10^2 Pa

(D) 9.2×10^4 Pa

25 A force F applied along the length of a wire produces a stress S. What would be the new stress if the load is reduced to $\frac{1}{4}$ of its value and the diameter of the wire is doubled?

(A) $\frac{S}{8}$

(B) $\frac{S}{8F}$

(C) $\frac{S}{16}$

(D) $16S$

Unit 2: Electricity and Magnetism, A.C. Theory and Electronics, Atomic and Nuclear Physics

Module 1: Electricity and Magnetism
2.1.1: Electrical Quantities

1 In which of the following equations does the left-hand side represent the energy transferred per unit time?

(A) $E = QV$
(B) $I = \dfrac{Q}{t}$
(C) $P = VI$
(D) $Q = CV$

2 Which of the following is NOT equivalent to the unit of resistivity?

(A) $C\,F^{-1}\,m\,A^{-1}$
(B) $\Omega\,m$
(C) $m\,A\,V^{-1}$
(D) $J\,m\,C^{-1}\,A^{-1}$

Items 3–4 refer to the following situation. A power of 4.0 W is used when a current of 0.20 A flows through a resistor for a period of 5.0 s.

3 The potential difference across the resistor and the energy used by it are respectively

(A) 20 V   4.0 J
(B) 0.8 V   4.0 J
(C) 20 V   20 J
(D) 4.0 V   20 J

4 The resistance of the resistor and the charge which flowed for the period are respectively

(A) 0.80 Ω   100 C
(B) 10 Ω   1.0 C
(C) 0.80 Ω   1.0 C
(D) 100 Ω   1.0 C

5 P and Q are resistance wires of the same material. The length of P is 3 times that of Q and the diameter of the cross-section of P is also 3 times that of Q. When connected in series to a power supply the total p.d. across P and Q is V_T. What is the ratio of the p.d. across P to the total p.d.?

(A) 1:3

(B) 1:4

(C) 1:6

(D) 1:9

Ⓐ Ⓑ Ⓒ Ⓓ

6 What do n and v respectively represent in the general equation $I = nAvq$ for the rate of flow of charge through a conductor?

(A) number of free charges   voltage

(B) number of free electrons per unit volume   velocity

(C) number of free charges   volume

(D) number of free charges per unit volume   velocity

Ⓐ Ⓑ Ⓒ Ⓓ

7 What is the length of a wire of resistivity 1.5×10^{-6} Ω m if its resistance is 10 Ω when its cross-sectional area is 1.5 mm^2?

(A) 0.10 m

(B) 1.5 m

(C) 10 m

(D) 15 m

Ⓐ Ⓑ Ⓒ Ⓓ

8 A cylindrical resistor has a resistance R. What would be its resistance if its length and diameter are both doubled?

(A) $\frac{R}{4}$ (C) 2R

(B) $\frac{R}{2}$ (D) 8R

Ⓐ Ⓑ Ⓒ Ⓓ

2.1.1: Electrical Quantities (cont.)

9 A potential difference of 8.0 V is set up across a wire of length 1.6 m, cross-sectional area 4.0 mm² and resistivity 2.0×10^{-6} Ω m. What is the current flowing through the wire?

(A) 10 A

(B) 2.5 A

(C) 1.0×10^{-4} A

(D) 1.0×10^{-1} A

10 A resistance wire X is connected across a battery of negligible internal resistance. X is replaced by a resistance wire Y of the same material and length but of twice the cross-sectional area. Which of the following is true?

(A) The current doubles and therefore the p.d. across the wire increases.

(B) The current is halved and therefore the p.d. across the wire decreases.

(C) The current and the p.d. across the wire remain the same.

(D) The current doubles but the p.d. across the wire remains the same.

11 A drift velocity of 0.40 mm s⁻¹ exists in a wire of diameter 2.0 mm. Determine the current flowing if the number of free electrons per unit volume is 2.5×10^{21} mm⁻³.

(A) 1.6×10^{3} A

(B) 5.0×10^{-9} A

(C) 5.0×10^{2} A

(D) 8.0 A

12 The p.d. across a wire is gradually increased with the temperature remaining constant. Which graph indicates how the drift velocity v of the electrons travelling in the wire will vary?

(A)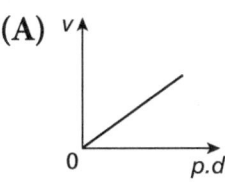

(B)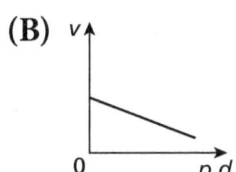

(C)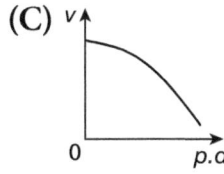

(D)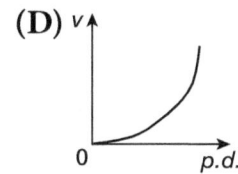

2.1.2: Electrical Circuits

1 Which of the following is true of Kirchhoff's laws?

I. The first law is a consequence of the conservation of energy and the second law a consequence of the conservation of charge.

II. The first law states that the algebraic sum of the currents entering a junction point is zero.

III. In traversing a closed loop of a circuit the total change in potential is zero.

(A) I only

(B) II only

(C) II and III only

(D) III only

2.1.2: Electrical Circuits (cont.)

Items **2–3** refer to the following graph of potential difference against current.

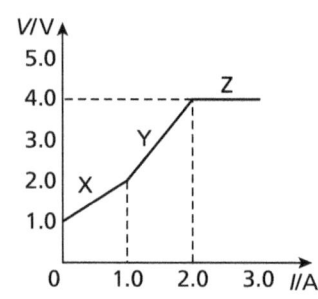

2 Where on the graph is Ohm's law obeyed?

(A) X

(B) Y

(C) Z

(D) X and Y

3 Which of the following is/are true?

   I. The resistance along X changes.

   II. The resistance along Y is 2.0 Ω.

   III. The resistance along Z is constant.

(A) I and II only

(B) II only

(C) II and III only

(D) III only

4 A cell of e.m.f. E and of internal resistance r supplies a current I to power a bulb of resistance R. What fraction of the power supplied does the bulb obtain?

(A) $\dfrac{R}{R+r}$

(B) $\dfrac{r}{R+r}$

(C) $\dfrac{R}{E}$

(D) $\dfrac{Ir}{E}$

5 A battery of e.m.f. 3.0 V drives a current of 0.50 A through an external resistance of 4.0 Ω. The internal resistance of the battery is

(A) 1.0 Ω

(B) 1.5 Ω

(C) 2.0 Ω

(D) 6.0 Ω

6 What are the values of I_1, I_2 and V in the following circuit?

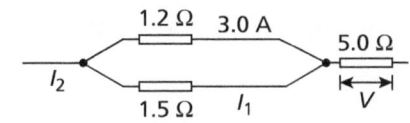

	I_1	I_2	V
(A)	2.4 A	5.4 A	27 V
(B)	1.8 A	4.8 A	24 V
(C)	2.7 A	5.7 A	13 V
(D)	3.8 A	6.8 A	34 V

Item 7 refers to the following circuit. The internal resistance of the cell is 1.0 Ω and the voltmeter is of infinite resistance.

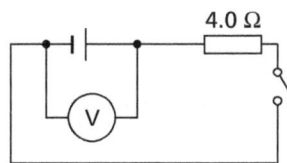

7 If the voltmeter reads 6.0 V when the switch is closed, what is the e.m.f. of the cell?

(A) 6.0 V

(B) 7.2 V

(C) 7.5 V

(D) 5.0 V

2.1.2: Electrical Circuits (cont.)

Items **8–9** refer to the following diagram. The cell has negligible internal resistance and the voltmeter has infinite resistance.

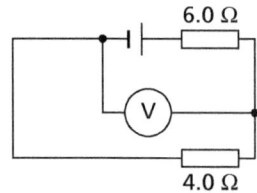

8 If the voltmeter reading is 4.0 V what is the current flowing in the circuit?

(A) 1.5 A

(B) 0.67 A

(C) 0.40 A

(D) 1.0 A

9 What is the e.m.f. of the cell?

(A) 6.0 V

(B) 10 V

(C) 4.0 V

(D) 5 V

10 In which component is the rate of energy consumption a minimum?

(A) 5 A, 1 Ω

(B) 2 A, 8 Ω

(C) 4 A, 2 V

(D) 20 Ω, 4 V

11 Each resistor in the following section of a circuit is of resistance 2.0 Ω.

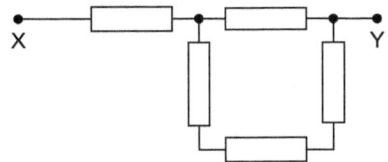

The equivalent resistance connected between X and Y is

(A) 3.0 Ω
(B) 3.5 Ω
(C) 4.0 Ω
(D) 10 Ω

Items **12–13** refer to the following Wheatstone bridge circuit. The rheostat is adjusted until there is no deflection on the galvanometer.

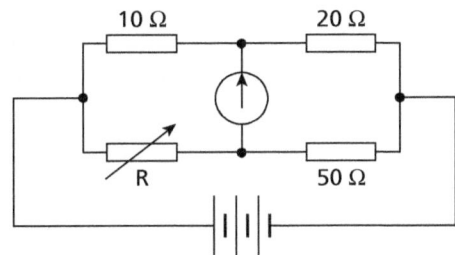

12 What is the resistance of the rheostat R?

(A) 100 Ω
(B) 25 Ω
(C) 4.0 Ω
(D) 0.25 Ω

13 If the e.m.f. of the battery is 3.0 V and it has negligible internal resistance, what current flows through the 20 Ω resistor?

(A) 0.15 A
(B) 0.067 A
(C) 0.10 A
(D) 0.14 A

2.1.2: Electrical Circuits (cont.)

Item 14 refers to the following circuit.

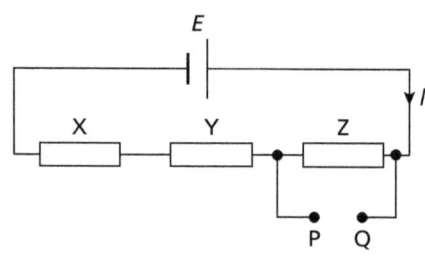

14 The cell is of e.m.f E and has negligible internal resistance. The resistances of X, Y and Z are R_X, R_Y and R_Z respectively. What is the voltage across the terminals PQ?

(A) $E\left(\dfrac{R_Z}{R_X + R_Y + R_Z}\right)$

(B) $E\left(\dfrac{R_X + R_Y + R_Z}{R_Z}\right)$

(C) $\dfrac{I}{E}(R_Z)$

(D) $\dfrac{I}{E}(R_X + R_Y + R_Z)$

Item 15 refers to the following circuit.

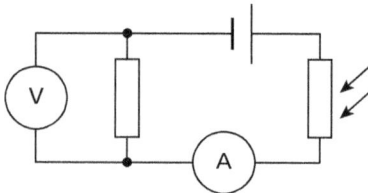

15 In bright light the resistance of the LDR decreases. How would the readings on the voltmeter and ammeter be affected?

	Voltmeter	Ammeter
(A)	increases	decreases
(B)	decreases	increases
(C)	increases	increases
(D)	unchanged	increases

Item 16 refers to the following circuit.

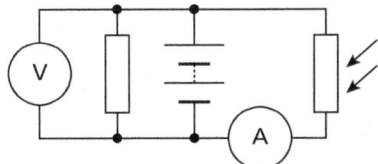

16 As it becomes dark the resistance of the LDR increases. How would the readings on the voltmeter and ammeter be affected?

	Voltmeter	**Ammeter**
(A)	decreases	increases
(B)	unchanged	decreases
(C)	unchanged	increases
(D)	increases	decreases

Item 17 refers to the following circuit, which contains a rheostat with a range of $0 \rightarrow 20\ \Omega$.

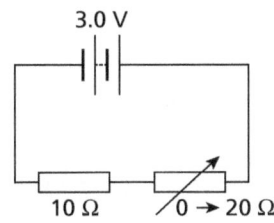

17 If the e.m.f. of the cell is 3.0 V what are the least and greatest values of the p.d. that can be obtained across the 10 Ω resistor?

	Least	**Greatest**
(A)	0 V	3.0 V
(B)	0 V	1.0 V
(C)	1.0 V	3.0 V
(D)	1.0 V	2.0 V

2.1.2: Electrical Circuits (cont.)

18 The cell in the following circuit is of e.m.f. 3 V and has negligible internal resistance. What are the readings on X, Y and Z?

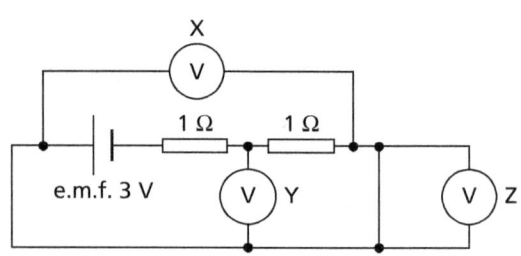

	X	Y	Z
(A)	3 V	1 V	1.5 V
(B)	3 V	0 V	0 V
(C)	0 V	2 V	3 V
(D)	0 V	1.5 V	0 V

Item 19 refers to the following circuit.

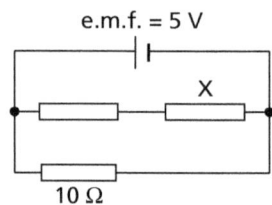

19 The current through X is 0.4 A. What is the current supplied by the cell if it has negligible internal resistance?

(A) 0.9 A

(B) 0.8 A

(C) 1.0 A

(D) 1.2 A

20 The battery of the following circuit has no internal resistance. What is the current through the 3.0 Ω resistor?

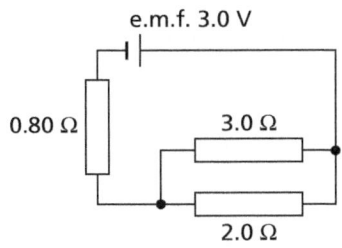

(A) 1.0 A

(B) 0.90 A

(C) 0.60 A

(D) 0.40 A

Ⓐ
Ⓑ
Ⓒ
Ⓓ

Items **21–23** refer to the following diagrams. The ammeter and voltmeter are assumed ideal and the cell has an internal resistance r. The graph shows the variation between the voltage V across the rheostat and the current I through it as the load resistance R is altered.

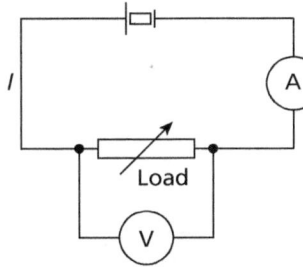
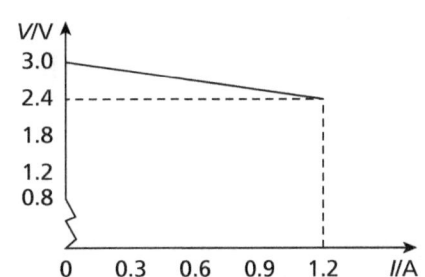

21 What is the e.m.f. E of the cell and its lost voltage V_L when delivering 1.2 A?

	E	V_L
(A)	2.4 V	1.2 V
(B)	3.0 V	0.6 V
(C)	3.0 V	2.4 V
(D)	2.4 V	0.6 V

Ⓐ
Ⓑ
Ⓒ
Ⓓ

2.1.2: Electrical Circuits (cont.)

22 When the voltmeter reads 2.4 V, the current is 1.2 A. What is the internal resistance of the cell?

(A) 2.0 Ω

(B) 1.5 Ω

(C) 1.0 Ω

(D) 0.5 Ω

23 Which of the following graphs shows how the terminal p.d. V varies as the load resistance R increases?

(A)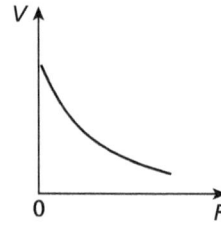

(C)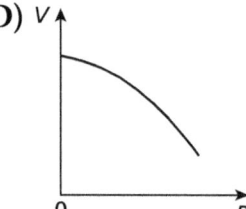

(B)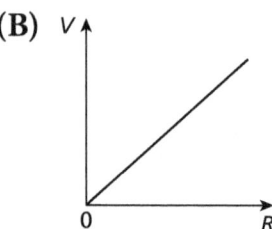

(D)

24 Which relation is true of the currents at the junction shown in the following diagram?

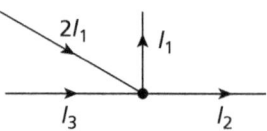

(A) $I_1 = I_2 + I_3$

(B) $I_1 = I_3 - I_2$

(C) $I_2 - I_1 = I_3$

(D) $I_1 + I_2 + I_3 = 0$

2.1.3: Electric Fields

1 The SI unit of electric field strength is

(A) N C

(B) V C^{-1}

(C) V m

(D) N C^{-1}

2 Permittivity can be expressed in

(A) C^2 m^{-2}

(B) C^2 m^{-2} N^{-1}

(C) A^2 s^2 m^{-2}

(D) N C^{-2} m^2

Item 3 refers to the following diagram. X and Y are horizontal metal plates.

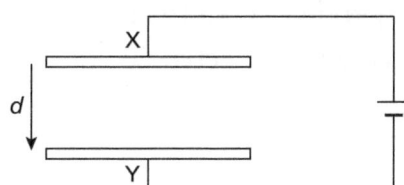

3 Which of the following graphs represents the relationship between the electric field strength E existing between the plates X and Y, and the distance d shown in the diagram?

(A)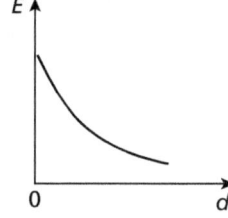

(C)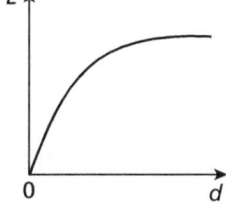

(B)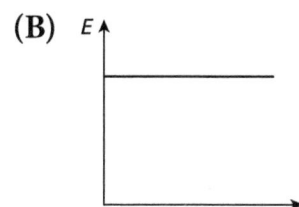

(D)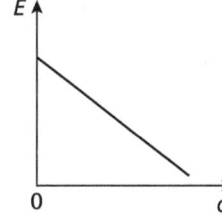

2.1.3: Electric Fields (cont.)

4 The electric potential 30 cm from a point charge in a vacuum is 3.0×10^3 V. What is the value of the charge?

(A) 1.2×10^6 C
(B) 4.0×10^{-6} C
(C) 1.0×10^{-7} C
(D) 1.3×10^{-6} C

5 What is the electric field strength at a distance 50 cm from a point charge of 5.0 μC in a vacuum?

(A) 1.8×10^5 N C^{-1}
(B) 9.0×10^4 N C^{-1}
(C) 1.8×10^1 N C^{-1}
(D) 2.3×10^6 N C^{-1}

6 Parallel plates 10 cm apart have a p.d. of 1.0 kV between them. An electron enters the region between the plates perpendicular to the field at a speed of 4.0×10^6 m s^{-1}. What is the value of the electric force on the electron?

(A) 1.6×10^{-15} N
(B) 4.0×10^{10} N
(C) 6.4×10^{-5} N
(D) 6.4×10^{-9} N

Item 7 refers to the following diagram which shows a subatomic particle moving at high speed through a field. The path of the particle is illustrated by the dashed line.

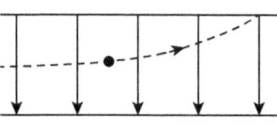

Field directed downward

7 What is the nature of the field and the possible type of particle?

	Nature of field	Possible type of particle
(A)	magnetic field	electron
(B)	electric field	proton
(C)	magnetic field	proton
(D)	electric field	electron

8 Which of the following is/are true of the electric field lines used to represent electric fields?

    I. They attract each other when they are close side-by-side.

    II. They appear to have a longitudinal tension within.

    III. They are directed from places of higher potential to places of lower potential.

(A) I and II only

(B) I and III only

(C) II and III only

(D) I, II and III

9 Which of the following statements is/are true?

    I. The electrical potential at a point is the work done in transferring positive charge from infinity to the point.

    II. The electrical potential difference between two points is the work done by an external agent in transferring unit positive charge between those points.

    III. The electrical potential at a point is the work done by an external agent in transferring unit positive charge from infinity to the point.

(A) I only

(B) I and III only

(C) II and III only

(D) III only

10 A pair of parallel plates 20.0 cm apart has a potential difference of 400 V between them. What is the electric field strength at a point between the plates which is 5.0 cm from the positive plate?

(A) 0.50×10^3 V m^{-1}

(B) 2.0×10^3 V m^{-1}

(C) 4.0×10^3 V m^{-1}

(D) 8.0×10^3 V m^{-1}

2.1.3: Electric Fields (cont.)

Items **11–13** refer to the following diagram.

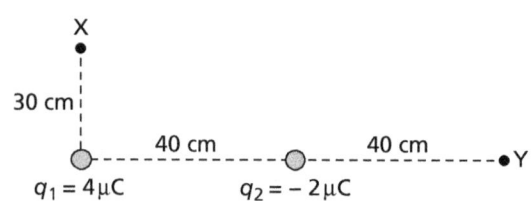

11 The electrical potential at point X due to q_1 and q_2 is

(A) 1.38×10^5 V

(B) 7.5×10^4 V

(C) 8.4×10^4 V

(D) 1.6×10^5 V

12 The electric field strength at point Y due to q_1 and q_2 is

(A) 1.7×10^5 N C^{-1} to the right

(B) 5.6×10^4 N C^{-1} to the left

(C) 1.7×10^5 N C^{-1} to the left

(D) 5.6×10^4 N C^{-1} to the right

13 The horizontal forces exerted on q_1 and q_2 are:

	Force on q_1	Force on q_2
(A)	0.45 N to the right	0.225 N to the left
(B)	0.225 N to the right	0.45 N to the left
(C)	0.45 N to the right	0.45 N to the left
(D)	0.45 N to the left	0.45 N to the right

14 The electric potential and electric field strength at a point 1 m away from a point charge q are respectively V and E. What would be the respective values of the electrical potential and electrical field strength at a point 2 m away?

(A) $\dfrac{V}{2}$ $\dfrac{E}{4}$

(B) $\dfrac{V}{4}$ $\dfrac{E}{2}$

(C) $\dfrac{V}{4}$ $\dfrac{E}{16}$

(D) V E

15 What is the electric field strength at a point 2 cm from the lower plate?

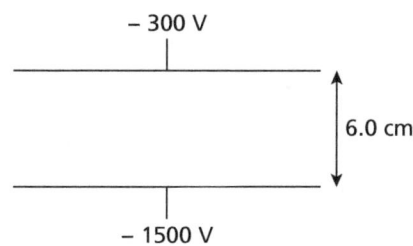

(A) 18 V m^{-1} directed down

(B) 20 V m^{-1} directed down

(C) 2.0×10^4 V m^{-1} directed down

(D) 20 V m^{-1} directed up

16 A very small positive test charge is placed in an electric field. The force per unit charge on it is the

(A) electrical flux density

(B) electrical potential energy

(C) electrical potential

(D) electrical field strength

17 Which of the following is/are true? Coulomb's law implies that the force between point charges is

    I. proportional to the masses of the charges.

    II. inversely proportional to the square of the distance between the charges.

    III. directly proportional to the permittivity of the medium in which the charges are immersed.

(A) I and II only

(B) II only

(C) I and III only

(D) II and III only

2.1.4: Capacitors

1 The SI unit of capacitance is equivalent to
- (A) $C\,J^{-1}$
- (B) $V\,C^{-1}$
- (C) $C\,V^{-1}$
- (D) $J\,C^{-1}$

2 Which of the following graphs concerning the energy E stored in a capacitor, the p.d. V across its plates and the charge Q it stores is NOT correct?

(A)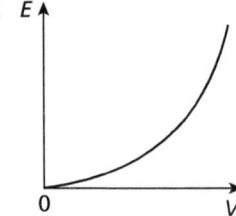

(C)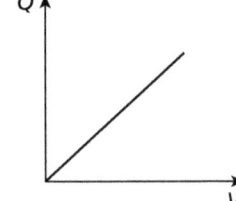

(B)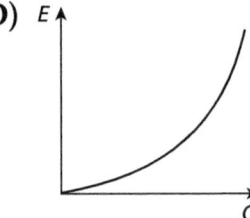

(D)

3 What is the energy stored in a capacitor of capacitance 200 μF when the p.d. between its plates is 500 V?
- (A) 25 J
- (B) 50 J
- (C) 2500 J
- (D) 5000 J

4 What is the energy stored in a capacitor of capacitance 200 μF when the charge stored on its plates is 5.0 μC?

(A) 6.3×10^{-8} J

(B) 1.3×10^{-2} J

(C) 2.5×10^{-2} J

(D) 1.3×10^{-7} J

5 The charge on a capacitor is 3.0 C when the p.d. between its plates is 500 V. What is the charge on the plates when the p.d. drops to 125 V?

(A) 0.75 C

(B) 1.25 C

(C) 2.5 C

(D) 3.0 C

6 A capacitor of capacitance C has circular plates of diameter d which are a distance x apart. What would be the new capacitance if d was doubled and x halved?

(A) $4C$

(B) $\frac{1}{4}C$

(C) $8C$

(D) $\frac{1}{8}C$

7 A capacitor of capacitance C has rectangular plates of area A, which are a distance d apart. Half of the air space between the plates is filled from one plate to the next with a material of relative permittivity 2. What is the new capacitance?

(A) $\frac{1}{2}C$ (C) $\frac{3}{2}C$

(B) C (D) $\frac{2}{3}C$

2.1.4: Capacitors (cont.)

8 In the following diagrams, each individual capacitor has the same capacitance. Which arrangement produces the LEAST total capacitance?

(A) ⊣⊢⊣⊢ (C) (parallel-series combination)

(B) (series-parallel combination) (D) ⊣⊢

Ⓐ Ⓑ Ⓒ Ⓓ

9 What is the total capacitance obtained between the terminals X and Y in the following diagram if each individual capacitor is of capacitance C?

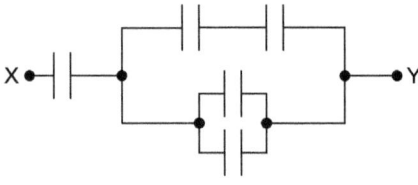

(A) $\dfrac{5}{7}C$ (C) $\dfrac{2}{7}C$

(B) $\dfrac{7}{5}C$ (D) $\dfrac{7}{2}C$

Ⓐ Ⓑ Ⓒ Ⓓ

10 Which of the following does NOT give the energy stored in a capacitor of capacitance C when a charge of Q exists on its plates due to a p.d. V?

(A) $\dfrac{1}{2}CV^2$ (C) $\dfrac{1}{2}\dfrac{V^2}{Q}$

(B) $\dfrac{1}{2}\dfrac{Q^2}{C}$ (D) $\dfrac{1}{2}QV$

Ⓐ Ⓑ Ⓒ Ⓓ

Items 11–14 refer to the following situation.

A capacitor of capacitance 4.0 µF has a p.d. of 10 V across its plates when fully charged. It discharges through a resistor of resistance 5.0×10^3 Ω.

11 How long will it take for the voltage across the plates to fall to 2.0 V?

- (A) 3.2×10^{-2} s
- (B) 4.0×10^{-3} s
- (C) 1.2×10^{-2} s
- (D) 2.0×10^{-2} s

Ⓐ Ⓑ Ⓒ Ⓓ

12 What is the initial charge on the plates just before discharge?

- (A) 2.0×10^{-3} C
- (B) 4.0×10^{-7} C
- (C) 4.0×10^{-5} C
- (D) 2.5×10^{5} C

Ⓐ Ⓑ Ⓒ Ⓓ

13 How long will it take for the charge to fall to ¼ of its initial value?

- (A) 5.0×10^{-3} s
- (B) 2.8×10^{-2} s
- (C) 5.8×10^{-3} s
- (D) 2.0×10^{-2} s

Ⓐ Ⓑ Ⓒ Ⓓ

14 What is the value of the time constant of the discharge circuit?

- (A) 0.01 s
- (B) 0.02 s
- (C) 0.03 s
- (D) 0.04 s

Ⓐ Ⓑ Ⓒ Ⓓ

2.1.4: Capacitors (cont.)

15 Which of the following is/are true in relation to capacitors of capacitance C discharging through a resistance R?

   I. The product RC is known as the time constant and it can be measured in seconds.

   II. The larger the time constant, the faster the capacitor will discharge.

   III. In a time equal to the time constant, the capacitor will discharge to about 37% of its initial charge.

(A) I only

(B) I and II only

(C) I and III only

(D) III only

2.1.5: Magnetic Fields and Forces

1 The tesla is equivalent to

(A) $J\,C^{-1}\,m^{-1}\,s$

(B) $N\,A^{-1}\,m^{-1}$

(C) $N\,C\,m^{-1}\,s$

(D) $V\,m^{-1}$

2 What is the magnetic field strength at a point 20 cm below a straight horizontal wire that carries a current of 2.0 A from west to east?

(A) 2.0×10^{-6} T   directed to the north

(B) 6.3×10^{-6} T   directed to the north

(C) 6.3×10^{-6} T   directed to the south

(D) 2.0×10^{-6} T   directed to the south

3 A short coil of 250 turns carrying a current of 8.0 A produces a magnetic field of strength 30 mT at its centre. What is the radius of the coil?

(A) 6.7×10^{-2} m

(B) 6.7×10^{-3} m

(C) 8.4×10^{-2} m

(D) 4.2×10^{-2} m

4 A coil is wrapped at 50 turns per cm of its length with copper wire. It is 40 cm long and carries a current of 2.0 A. What is the magnetic field strength at its centre along its axis?

(A) 5.0×10^{-3} T

(B) 1.3×10^{-4} T

(C) 1.3×10^{-2} T

(D) 5.0×10^{-5} T

5 A solenoid of N turns is of length L. When a current I flows in the coil a magnetic field of 50 μT is created along its axis. If the current is doubled and the number of turns is halved, what would be the resulting magnetic field strength along the axis?

(A) 50 μT

(B) 200 μT

(C) 12.5 μT

(D) 25 μT

6 The following diagram shows four cylindrical coils carrying currents. The broken lines indicate the directions of magnetic fields. In which coil is the magnetic field direction correctly shown?

(A)

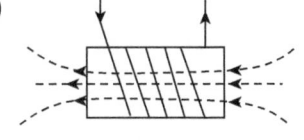

(B)

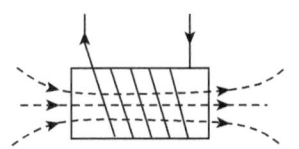

(C)

(D)

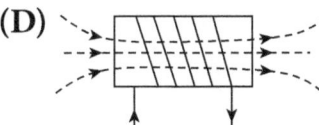

2.1.5: Magnetic Fields and Forces (cont.)

7 In which of the following diagrams are the forces *F* correctly directed on the wires which carry currents *I*?

(A)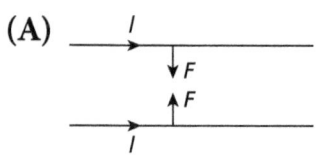

(C)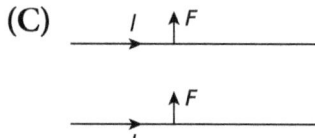

(B)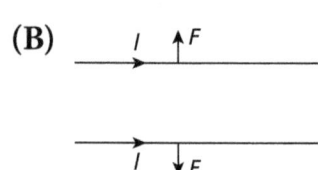

(D)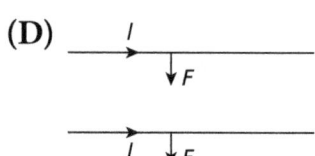

Ⓐ
Ⓑ
Ⓒ
Ⓓ

8 X, Y and Z show wires carrying currents through magnetic fields. In X, the current flows into the plane of the paper, and in Y and Z, it flows in the direction indicated by the arrows. Which of the following is true for the direction of the forces created on the wires of diagrams X, Y and Z?

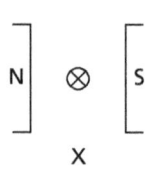 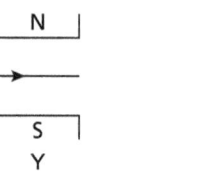

⊗ = Current into plane of paper

	X	Y	Z	
(A)	to top of page	into page	out of page	Ⓐ
(B)	to top of page	into page	out of page	Ⓑ
(C)	to bottom of page	into page	into page	Ⓒ
(D)	to the left	to top of page	to the right	Ⓓ

9 The following diagram shows an electron beam entering a uniform magnetic field that is perpendicular to its path. The crosses in the diagram represent the magnetic field acting into the plane of the paper. In which direction will the beam swerve?

× × × ×
× × × ×   × Represents a magnetic
Electron   field perpendicularly
beam × × × ×   into the plane of the
× × × ×   diagram.

(A) To the top of the page.   Ⓐ

(B) To the bottom of the page.   Ⓑ

(C) Out of the page.   Ⓒ

(D) Into the page.   Ⓓ

10 Two ions, X and Y, are of the same charge, but X is three times the mass of Y. The particles have the same speed and direction when they enter a uniform magnetic field which is perpendicular to their paths. They both take on circular paths. If the radius of the path of X is R, what is the radius of the path of Y?

(A) $\dfrac{R}{3}$ (C) $\dfrac{3}{R}$

(B) $3R$ (D) $\dfrac{9}{R}$

Ⓐ Ⓑ Ⓒ Ⓓ

11 Two ions, X and Y, are of the same mass, but X has three times the charge of Y. The particles are travelling at the same speed and direction when they enter a uniform magnetic field which is perpendicular to their paths. They both take on circular paths. If the radius of the path of X is R, what is the radius of the path of Y?

(A) $\dfrac{R}{3}$ (C) $\dfrac{3}{R}$

(B) $3R$ (D) $\dfrac{9}{R}$

Ⓐ Ⓑ Ⓒ Ⓓ

12 The following diagram shows sections of four current-carrying conductors A, B, C and D placed in a magnetic field directed to the right. On which is the electromagnetic force greatest?

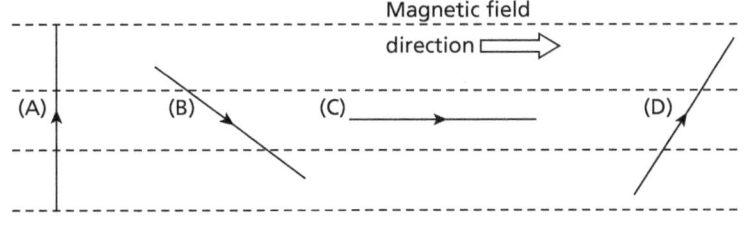

2.1.5: Magnetic Fields and Forces (cont.)

Item 13 refers to the following diagram. X and Y are wires carrying equal currents perpendicularly into the plane of the paper.

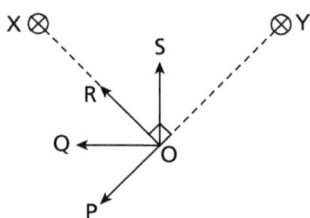

13. Which arrow BEST represents the direction of the resultant magnetic field produced at point O?

(A) P
(B) Q
(C) R
(D) S

Item 14 refers to the following diagram. X is a wire carrying a current perpendicularly into the plane of the paper.

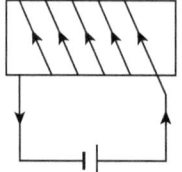 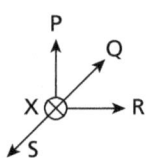

14. Which arrow BEST represents the direction of the electromagnetic force on X when current flows in the coil as shown?

(A) P
(B) Q
(C) R
(D) S

15 A magnetic field of strength 1.5 T is directed to the right at an angle of 30° below the horizontal. What is the magnitude of the force on a horizontal wire of length 25 cm that carries a current of 5.0 mA directed to the right?

(A) 9.4×10^{-2} N
(B) 18×10^{-3} N
(C) 9.4×10^{-4} N
(D) 1.8×10^{-1} N

Item 16 refers to the following diagram where three wires P, Q and R carry currents of the same magnitude. The distance between adjacent wires is the same.

16 The resultant force on each of P, Q and R, is directed

	P	Q	R
(A)	downward	upward	downward
(B)	upward	downward	upward
(C)	downward	downward	upward
(D)	upward	downward	downward

17 A current of 5.0 A flows from west to east along a horizontal wire of length 30 cm. A magnetic field of flux density 50 mT is directed perpendicularly to the wire and keeps it suspended. What is the mass of the wire and the direction of the magnetic field?

	Mass of wire	Direction of magnetic field
(A)	7.6 g	north
(B)	75 g	north
(C)	7.7 g	south
(D)	75 g	north

2.1.5: Magnetic Fields and Forces (cont.)

Item **18** refers to the following diagram.

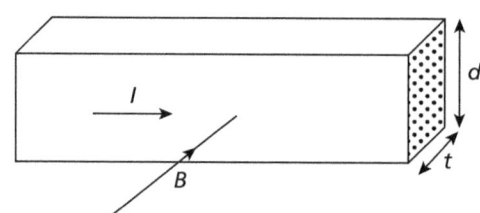

A current *I* of 8.0 A flows through a rectangular section of an n-type semiconductor, which has its majority charge carriers as electrons. Applying a magnetic field of flux density *B* of 1.0 T perpendicularly to the section as shown produces a Hall voltage of 5.0 mV.

18 Calculate the number of charge carriers per unit volume given that the lengths *t* and *d* are 1.2 cm and 1.0 cm respectively.

(A) 8.3×10^{23} m^{-3}

(B) 1.2×10^{20} m^{-3}

(C) 8.3×10^{17} m^{-3}

(D) 1.2×10^{14} m^{-3}

19 Electrons travelling horizontally at a velocity of 2.0×10^6 m s^{-1} enter a region between two horizontal plates 20 cm apart with a p.d. of 400 V across them. A horizontal magnetic field is directed perpendicularly through the region between the plates, allowing the electron beam to maintain a straight horizontal path. What is the magnetic flux density of the magnetic field?

(A) 4.0×10^{-5} T

(B) 1.0×10^{-3} T

(C) 4.0×10^{9} T

(D) 1.6×10^{8} T

2.1.6: Electromagnetic Induction

1 The SI units of magnetic flux density and magnetic flux are respectively:

(A) T, H

(B) H, Wb

(C) Wb, T

(D) T, Wb

2 A conducting coil of 50 turns has a diameter of 20 cm. What is the flux linkage with the coil when a magnetic field of strength 250 mT is directed perpendicularly through its circular face?

(A) 0.39 Wb

(B) 7.9×10^{-3} Wb

(C) 390 Wb

(D) 0.125 Wb

<u>Item 3</u> refers to the following diagram. A conducting rod XY of length 20 cm and diameter 2.0 mm moves east horizontally at 4.0 m s^{-1} through a magnetic field of flux density 5.0 mT directed at an angle of 30° below the horizontal.

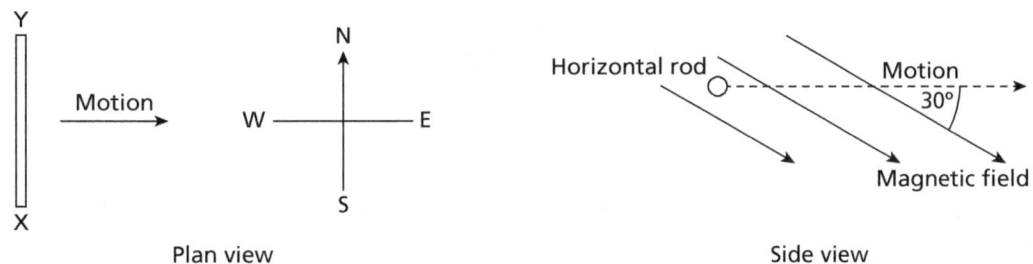

Plan view   Side view

3 What is the e.m.f. generated across the ends of the rod and which end is at a positive potential?

	e.m.f.	Positive end of rod
(A)	4.0×10^{-6} V	Y
(B)	4.0×10^{-6} V	X
(C)	2.0×10^{-3} V	Y
(D)	2.0×10^{-3} V	X

2.1.6: Electromagnetic Induction (cont.)

<u>Item 4</u> refers to the following diagram where a light aluminium ring stands with its axis aligned with that of an adjacent coil wrapped on a soft-iron core.

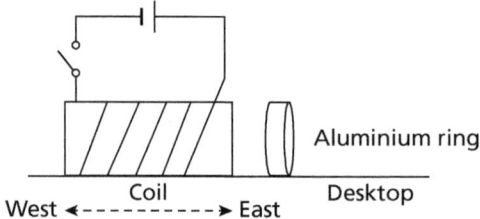

4 At the instant the switch is closed, the light aluminium ring will

(A) move to the west, towards the coil.

(B) move to the east, away from the coil.

(C) roll towards north.

(D) roll towards south.

<u>Item 5</u> refers to the following diagram. The aluminium rings X and Y stand at rest when a steady current flows in the coil.

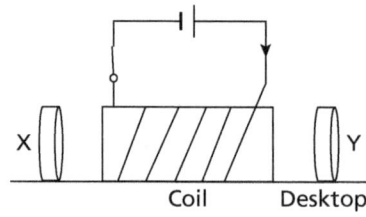

5 What is the effect on X and Y due to the current in the coil at the instant when the switch is opened?

(A) X is pushed to the left and Y is pushed to the right.

(B) X and Y are unaffected.

(C) X is pulled to the right and Y is pulled to the left.

(D) X and Y are respectively pulled and pushed to the right.

6 A bar magnet is allowed to fall through a long hollow vertical coil which is connected to a centre-zero galvanometer. Which graph BEST describes the current induced in the coil?

(A)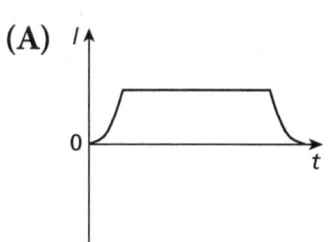

(C)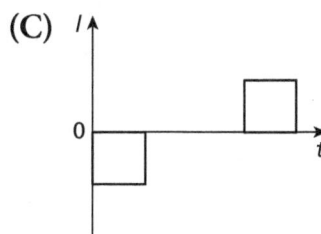

(B)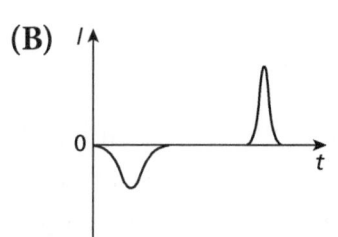

(D)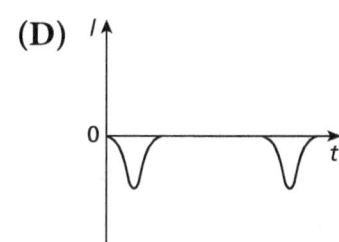

Ⓐ
Ⓑ
Ⓒ
Ⓓ

7 The core of a transformer is laminated with its laminations insulated because

(A) this strengthens the magnetic field within it.   Ⓐ

(B) the insulation between the laminations prevents electrical shock.   Ⓑ

(C) eddy currents induced within the core will be reduced significantly.   Ⓒ

(D) this allows the current through the core to have a more direct path.   Ⓓ

8 An ideal transformer has 100 turns on its secondary coil. It is used to operate a 6.0 V, 3.0 W device when its primary coil is connected to a 120 V a.c. supply. What is the current in the primary coil and how many turns does it have?

(A) 25 mA   2000 turns   Ⓐ

(B) 10 A   2000 turns   Ⓑ

(C) 0.5 A   20 turns   Ⓒ

(D) 0.025 A   200 turns   Ⓓ

2.1.6: Electromagnetic Induction (cont.)

9) A transformer steps down a voltage from 120 V to 30 V. If it is 80% efficient, what is the current in the primary coil when the secondary is connected to a resistance of 20 Ω?

(A) 4.8 A

(B) 6.0 A

(C) 0.47 A

(D) 2.1 A

10) A narrow coil of 5 turns and area 2.0×10^2 cm² is placed with its plane perpendicular to a uniform magnetic field of flux density 2.0×10^{-2} T. What is the average e.m.f. induced in the coil as it is pulled out of the field, in a direction parallel to its plane in a time of 0.25 s?

(A) 80 V

(B) 8.0×10^{-3} V

(C) 51 V

(D) 5.1×10^{-3} V

Items **11–13** refer to the following diagram, which shows two positions, X and Y, of a rectangular conducting coil of N turns and cross-sectional area A rotating at angular frequency ω within a magnetic field of flux density B. The axis of rotation is perpendicular to the field and to the page on which the diagram is drawn.

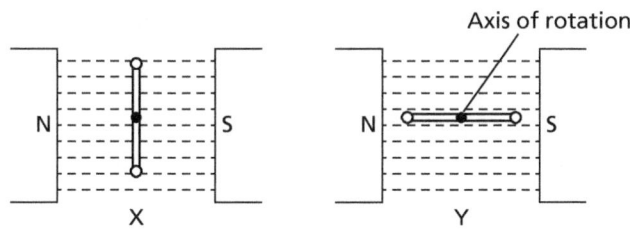

11 What is the maximum e.m.f. induced in the coil if the period of rotation is T and the angular frequency is ω?

(A) $NBA\omega \sin \dfrac{2\pi}{T}$

(B) $NBA\omega$

(C) NBA

(D) $NBA \sin \dfrac{2\pi}{T}$

12 What is the magnitude of the instantaneous flux linkage and e.m.f. respectively at any time t, if $t = 0$ when the coil is in position X?

	Flux linkage	**e.m.f.**
(A)	0	$NBA \sin \omega t$
(B)	$NBA \cos \omega t$	0
(C)	$NBA \cos \omega t$	$NBA\omega \sin \omega t$
(D)	$NBA \sin \omega t$	$NBA\omega \cos \omega t$

13 Which of the following graphs BEST describes the variation of e.m.f. E with time t, if $t = 0$ when in position Y?

(A)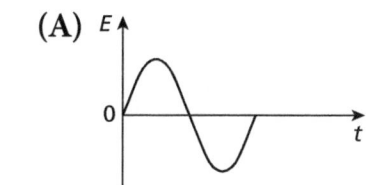

(C)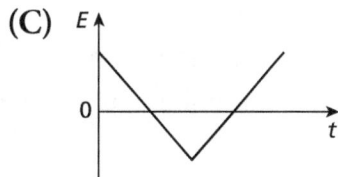

(B)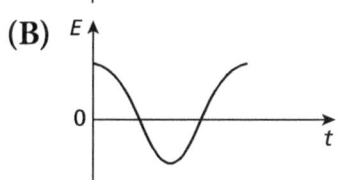

(D)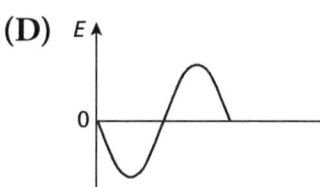

Module 2: A.C. Theory and Electronics
2.2.1: Alternating Currents

1 What is the r.m.s. value of an alternating voltage that has a peak value of 156 V?

(A) 220 V
(B) 78 V
(C) 110 V
(D) 312 V

Item 2 refers to the following graph, which shows the variation of voltage with time for an alternating voltage.

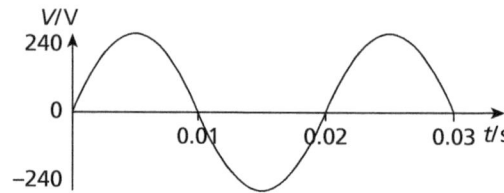

2 The instantaneous voltage is BEST represented by

(A) $V = 240 \sin(100\pi t)$
(B) $V = 480 \sin(100\pi t)$
(C) $V = 240 \sin(0.02\pi t)$
(D) $V = 240 \sin(0.01\pi t)$

3 A sinusoidal alternating voltage is connected across a resistor of 20 Ω. What is the power it delivers if the peak voltage is 170 V?

(A) 720 W
(B) 6.0 W
(C) 1.4×10^3 W
(D) 8.5 W

4 A sinusoidal alternating current with a peak-to-peak value of 14 A has an r.m.s value of

(A) 9.9 A

(B) 4.9 A

(C) 7.0 A

(D) 3.5 A

5 A power of 500 W is consumed by a resistor when a sinusoidal alternating current with peak value of 5.0 A flows through it. What is the resistance of the resistor?

(A) 40 Ω

(B) 316 Ω

(C) 1250 Ω

(D) 160 Ω

6 What is the peak-to-peak value and the frequency of the alternating current represented by the equation $I = 17 \sin(100\pi t)$? (I is measured in A and t is measured in seconds.)

(A) 34 A   100 Hz

(B) 24 A   50 Hz

(C) 34 A   50 Hz

(D) 24 A   100 Hz

7 A power P is consumed by a resistor R when a d.c. voltage V_x is placed across it. A sinusoidal voltage of r.m.s. value V_y is placed across a resistor of resistance $\frac{1}{8}R$ and the same power is consumed. The voltage V_y is

(A) $\dfrac{V_x}{2\sqrt{2}}$

(B) $\dfrac{V_x}{2}$

(C) $V_x\sqrt{8}$

(D) $V_x\sqrt{2}$

2.2.1: Alternating Currents (cont.)

8 A power P is consumed by a resistor R when a direct current I flows through it. A sinusoidal alternating current flows through R and consumes the same power. The r.m.s. value of the alternating current is

(A) $\dfrac{I}{\sqrt{2}}$

(B) $\dfrac{\sqrt{2}I}{2}$

(C) $I\sqrt{2}$

(D) I

9 A sinusoidal alternating supply of peak voltage 110 V produces a peak current of 5.0 A when placed across a device. What is the mean power consumed by the device?

(A) 390 W

(B) 275 W

(C) 550 W

(D) 780 W

Items **10–11** refer to the following situation.

The output V of an a.c. supply given by $V = 60 \sin(50\pi t)$ is connected across a resistance of 20 Ω.

10 What is the instantaneous power delivered at $t = 6.0$ ms?

(A) 2.0 W

(B) 118 W

(C) 4.9×10^{-2} W

(D) 20 W

11 What is the peak power and the mean power delivered?

	Peak power	Mean power
(A)	180 W	90 W
(B)	3.6 kW	1.8 kW
(C)	180 W	130 W
(D)	3.6 kW	2.5 kW

12 Which of the following is NOT an advantage of producing electrical energy as an alternating current to be transmitted to the consumer over long distances?

(A) Transformers may be used to step down the current so that less energy will be lost as heat in the transmitting cables. (A)

(B) Thinner and therefore less costly cable can be used to carry the smaller current stepped down by a transformer. (B)

(C) Wires offer less resistance to alternating currents than to direct currents. (C)

(D) The consumer can step up or down the voltage with minimum energy loss to be compatible with various devices. (D)

2.2.2: The p-n Junction Diode and Transducers

Item 1 refers to the following diagram.

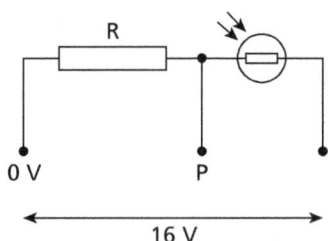

1 In bright sunlight the resistance of the LDR is 100 Ω, but in the dark it is 8.0 kΩ. At midday the potential at P is 12 V, but in the night it is approximately zero. What is the resistance of R?

(A) 1.2 kΩ (A)

(B) 300 Ω (B)

(C) 33 Ω (C)

(D) 6.0 kΩ (D)

2.2.2: The p-n Junction Diode and Transducers (cont.)

2 The LDR shown has infinite resistance in the dark. The voltmeter placed across it gives a reading of 12 V in the dark and just 2.0 V in bright sunlight. What is the resistance of R if the ammeter reads 4.0 mA in the day? (The voltmeter and ammeter are assumed ideal.)

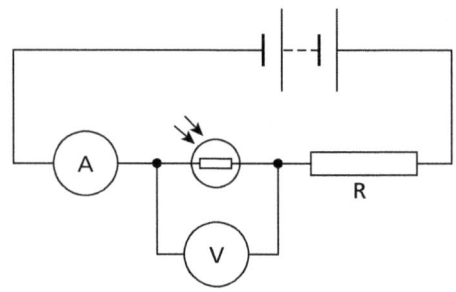

(A) 3.0 Ω

(B) 500 Ω

(C) 3.0 kΩ

(D) 2.5 kΩ

Ⓐ
Ⓑ
Ⓒ
Ⓓ

Item 3 refers to the following diagram.

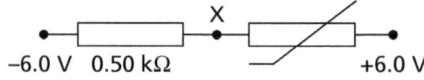

3 When the thermistor is at a certain high temperature T, its resistance is 0.10 kΩ. However, when in the cold, its resistance rises to more than ten times as much. What happens to the potential at X when the temperature drops significantly from T?

(A) Changes from +4.0 V to a negative potential

(B) Changes from −4.0 V to a positive potential

(C) Changes from +4.0 V to 0 V

(D) Changes from −4.0 V to 0 V

Ⓐ
Ⓑ
Ⓒ
Ⓓ

Item 4 refers to the following diagram.

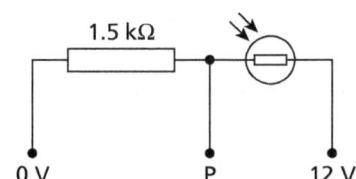

4. The LDR has a resistance of 1.0 kΩ in the dark, but at sunrise its resistance becomes just 100 Ω. How does the potential at P change at sunrise?

   (A) from 7.2 V to about 0.8 V
   (B) from 7.2 V to about 11 V
   (C) from 4.8 V to about 11 V
   (D) from 4.8 V to about 0.8 V

5. Which of the following circuits produces full-wave rectification of an a.c. input?

133

2.2.2: The p-n Junction Diode and Transducers (cont.)

6 In which of the following circuits is the rectified output being smoothed?

(A)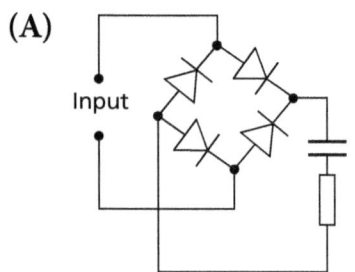

(C)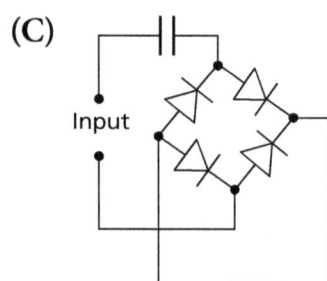

(B)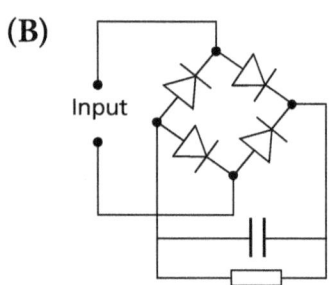

(D)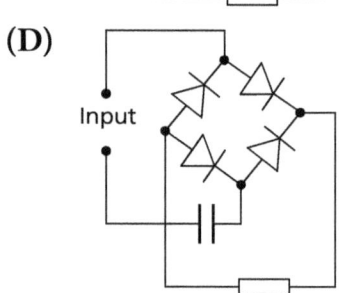

Item 7 refers to the following pair of diagrams. The diode X in circuit 1 is faulty and does not conduct in any direction.

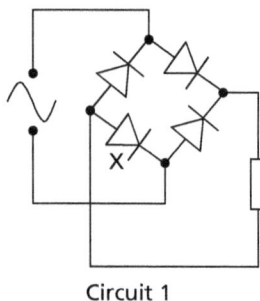
Circuit 1

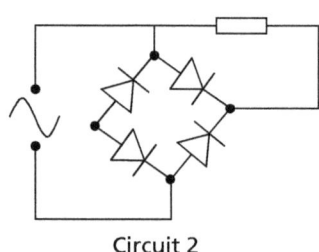
Circuit 2

7 Which pair of graphs BEST represents the variation of p.d. V across the loads with time t?

(A)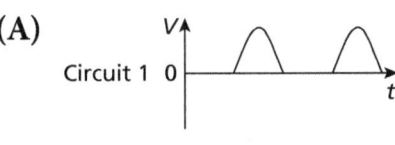

(C)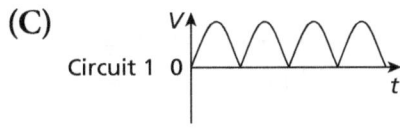

(B)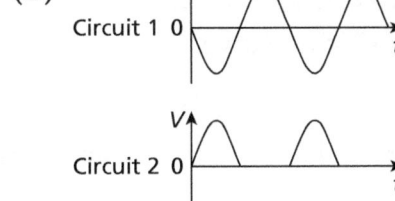

(D)

Ⓐ Ⓑ Ⓒ Ⓓ

Item 8 refers to the following circuits in which the input is a sinusoidal a.c.

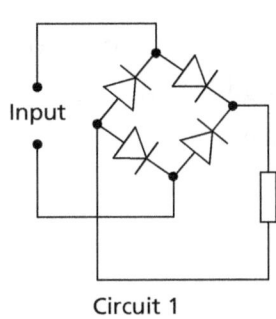

Circuit 1   Circuit 2

8 Which pair of graphs shows how the p.d. V across the load varies with time t for the two circuits?

(A)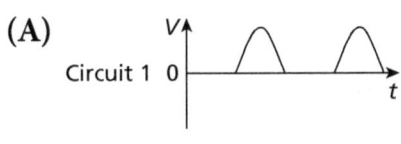

(C)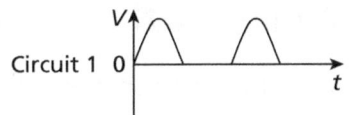

(B)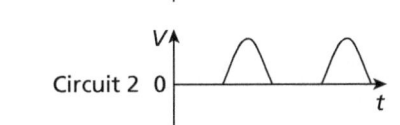

(D)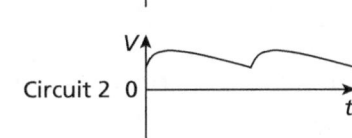

Ⓐ Ⓑ Ⓒ Ⓓ

2.2.2: The p-n Junction Diode and Transducers (cont.)

9 What is the effect on the depletion layer of a p-n junction when a p-n junction diode is connected in forward bias?

(A) It increases in width.

(B) It is unaffected.

(C) It decreases in width.

(D) Electrons cross the junction to the n-type material from the p-type material.

10 Which of the following is/are true of semiconducting materials?

   I. The majority charge carriers in a p-type material are holes.

   II. Both p-type and n-type materials have minority charge carriers which are electron-hole pairs that are thermally generated.

   III. P-type materials are formed when a small amount of an element with atoms that have three valence electrons is added to an element such as silicon, with atoms that have four valence electrons.

(A) I only

(B) I and II only

(C) II and III only

(D) I, II and III

11 Which of the following is NOT true?

(A) 'Drift currents' flow from places of higher electrical potential to places of lower electrical potential.

(B) 'Diffusion currents' are those that flow from places of higher charge density to places of lower charge density.

(C) Before being placed in a circuit, a p-n junction has an excess of electrons on the n-type side of the junction, which makes that side negative.

(D) When a p-n junction diode is operated in reverse bias at a p.d. less than the breakdown p.d., only the thermally generated minority charge carriers can cross the junction and produce a current.

12 Which of the following is NOT true of semiconducting diodes?

(A) LEDs are normally operated in forward bias and emit light when electrons jump into holes.

(B) Photodiodes are connected in reverse bias, so that they only conduct when the light intensity is high enough to generate minority charge carriers of electron-hole pairs.

(C) A Zener diode is operated in reverse bias at a potential high enough to cause an avalanche of electron-hole pairs to be produced.

(D) A Zener diode cannot be used as a voltage regulator.

13 Which of the following shows the correct circuit symbol for the transistor below it?

(A)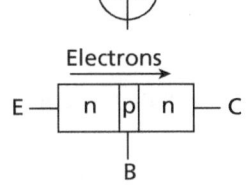

(C)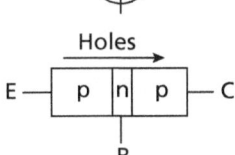

(B)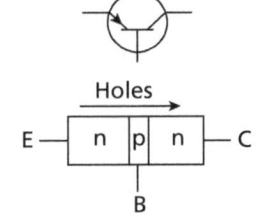

(D)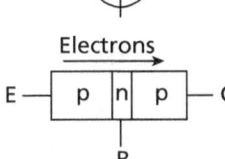

14 Which of the following is NOT true of transducers?

(A) A transducer is a device that receives a signal in one form of energy and converts it into a signal in another form of energy.

(B) Examples of input transducers are thermistors, light-dependent resistors and microphones.

(C) A sensor is an output transducer.

(D) Examples of output transducers are light-emitting diodes, buzzers and relays.

2.2.3: Operational Amplifiers

1 Which of the following is NOT true of operational amplifiers?

(A) An ideal operational amplifier has infinite input impedance, zero output impedance and infinite open loop gain.

(B) A practical operational amplifier has an open loop gain of the order of 10^5 for d.c. and low input frequencies. The gain decreases as the frequency increases.

(C) Energy enters the ideal operational amplifier at its inverting and non-inverting terminals.

(D) When used as a comparator, the output of the operational amplifier is saturated.

2 Which of the following is NOT true in relation to the gain and bandwidth of an operational amplifier?

(A) The bandwidth is the range of frequencies over which the gain is constant.

(B) Increasing the gain increases the bandwidth.

(C) Negative feedback decreases the gain.

(D) Reducing the gain reduces the chances of saturation of the output and can allow the system to be used as an analogue device.

Items 3–5 refer to the following circuit diagram of an operational amplifier with an open loop gain of 1.0×10^5.

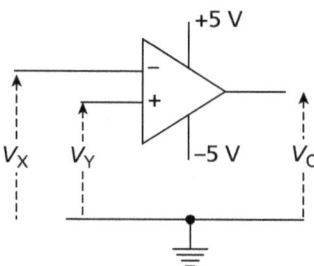

3 If $V_X = 40$ μV and $V_Y = 100$ μV, what is the voltage V_O produced at the output?

(A) 5 V

(B) 6 V

(C) 60 μV

(D) 50 μV

138

4 What is the MAXIMUM input voltage swing which will produce an unsaturated output?

(A) ±10 V

(B) ±50 μV

(C) ±5 V

(D) ±5 μV

5 If $V_X = -100$ μV and $V_Y = -80$ μV, what is the voltage V_O produced at the output?

(A) +2 V

(B) −2 V

(C) −18 V

(D) 5 V

6 Which of the following indicates when the LED is lit?

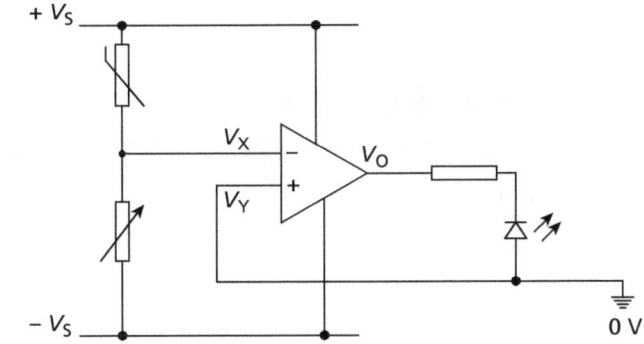

(A) $V_X > V_Y$

(B) $V_Y > V_X$

(C) $V_X = V_Y$

(D) $V_O > V_X - V_Y$

2.2.3: Operational Amplifiers (cont.)

7 What is the MINIMUM input voltage that can cause saturation in the following circuit if the open loop gain is 4×10^5?

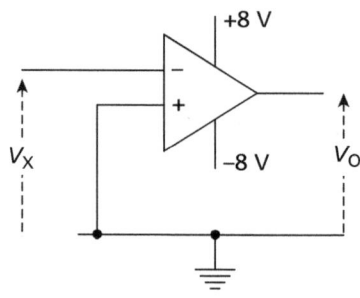

(A) $\pm 8 \ \mu V$

(B) $\pm 16 \ \mu V$

(C) $\pm 20 \ \mu V$

(D) $\pm 40 \ \mu V$

Ⓐ

Ⓑ

Ⓒ

Ⓓ

<u>Items 8–9</u> refer to the following circuit diagram and graphs. The variation of input voltage V_2 with time t is shown together with four graphs of the variation of output voltage V_O with time for the same period.

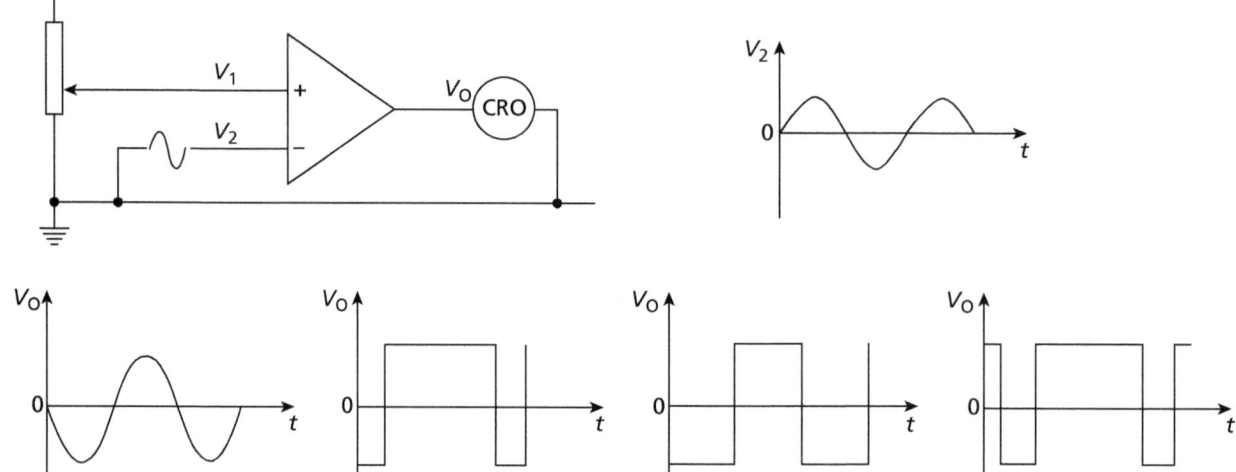

8 If $V_1 = 1$ V and V_2 is a sinusoidal a.c. of peak value 2 V, which of the graphs W, X, Y and Z BEST represents how V_O varies with time t?

(A) W

(B) X

(C) Y

(D) Z

9 The variable resistor is adjusted so that V_1 becomes 0 V. Which of the graphs W, X, Y and Z BEST represents how V_O now varies with time t?

(A) W

(B) X

(C) Y

(D) Z

Items **10–11** refer to the following circuit.

As night falls, the resistance of the LDR changes from less than 10 kΩ to greater than 10 kΩ.

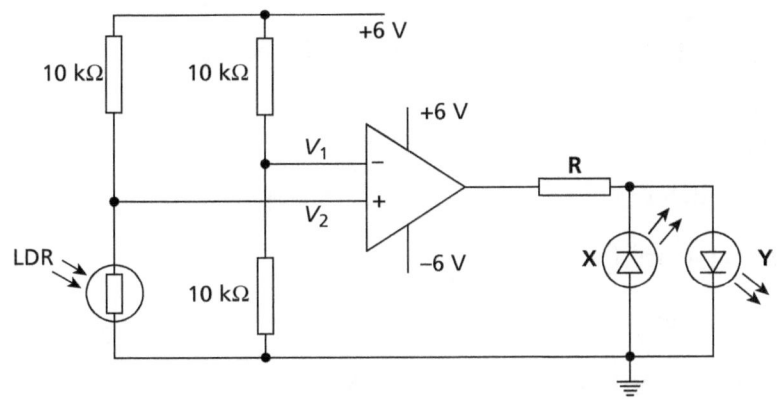

10 Which of the LEDs is lit during the day and which is lit during the night?

	Day	Night
(A)	X	Y
(B)	none	Y
(C)	X	none
(D)	Y	X

2.2.3: Operational Amplifiers (cont.)

11 The LEDs are rated at 1.2 V 30 mA. What would be a suitable value for the protective resistor R?

(A) 160 Ω

(B) 200 Ω

(C) 40 Ω

(D) 320 Ω

Items **12–16** refer to the following circuit diagram.

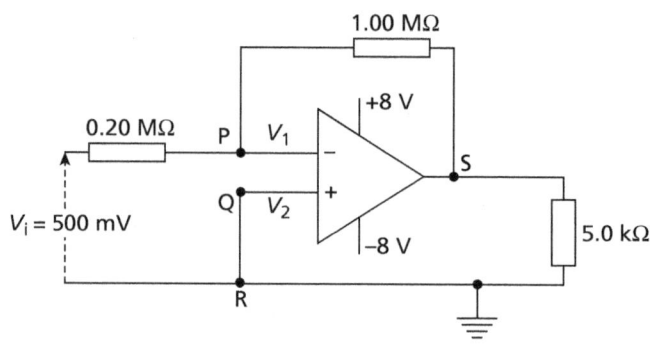

12 Give the name of the type of amplifier and state its gain.

	Type	Gain
(A)	Inverting amplifier	5
(B)	Non-inverting amplifier	6
(C)	Inverting amplifier	−5
(D)	Non-inverting amplifier	−6

13 What is the output voltage, and which point is known as virtual earth?

	Output voltage	Virtual earth
(A)	2.5 V	P
(B)	−2.5 V	P
(C)	3.0 V	Q
(D)	−3.0 V	Q

14 What is the output current and the current in the feedback resistor?

	Output current	Feedback current
(A)	0.50 mA	2.1 μA
(B)	0.50 mA	2.5 μA
(C)	0.60 mA	1.7 mA
(D)	0.60 mA	2.5 mA

15 If the input voltage was 2 V, what would be the output voltage?

(A) +10 V

(B) +8 V

(C) −8 V

(D) −10 V

16 What are the directions of current in the feedback resistor and in the input resistor?

	Feedback resistor	Input resistor
(A)	left to right	right to left
(B)	right to left	left to right
(C)	left to right	left to right
(D)	right to left	right to left

2.2.3: Operational Amplifiers (cont.)

Items **17–18** refer to the following diagram, where R_X, R_Y and R_Z are the resistances of the resistors X, Y and Z.

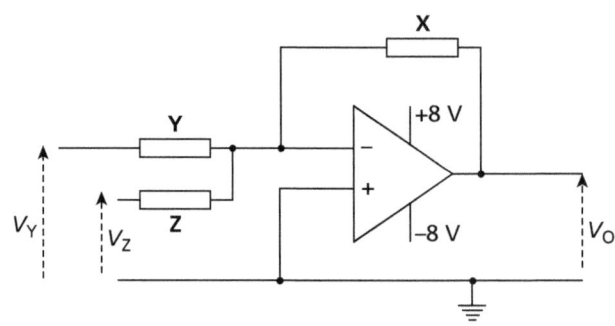

17 The output voltage V_O is given by the equation: $V_O = -(4V_Y + 2V_Z)$. Which selection of the following resistances would produce the required output?

	R_X/kΩ	R_Y/kΩ	R_Z/kΩ
(A)	1000	200	400
(B)	500	100	250
(C)	800	200	400
(D)	500	250	125

18 What is the value of V_O when V_Y and V_Z are 2 V and 1 V respectively?

(A) −8 V

(B) 8 V

(C) −10 V

(D) 10 V

Items **19–20** refer to the following diagrams. V_i is a sinusoidal alternating voltage of peak value 4 V and frequency 500 Hz.

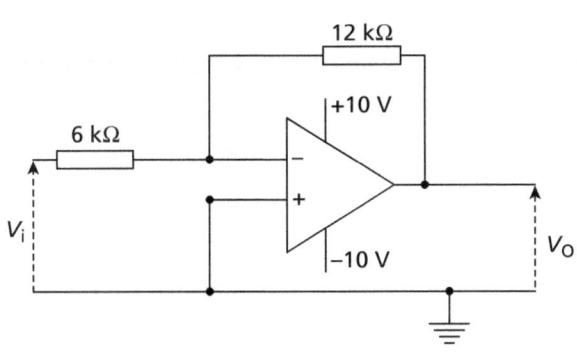
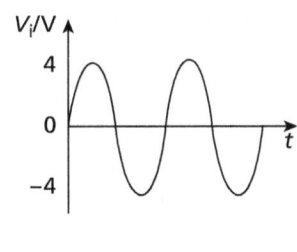

19 Which of the following graphs BEST indicates the output voltage V_O?

(A)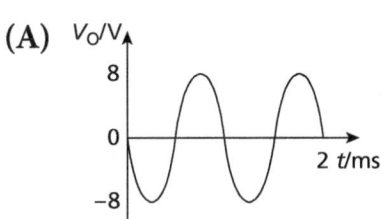

(C)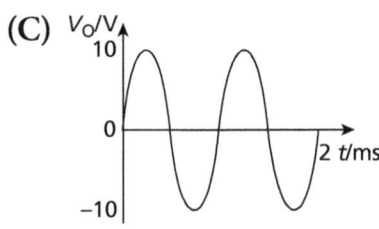

(B)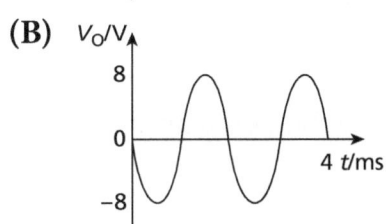

(D)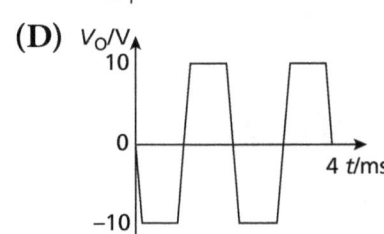

20 The peak value of the input voltage is changed to 8 V. Which of the following graphs BEST represents V_O?

(A)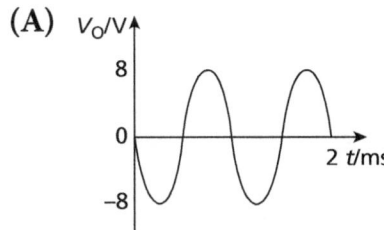

(C)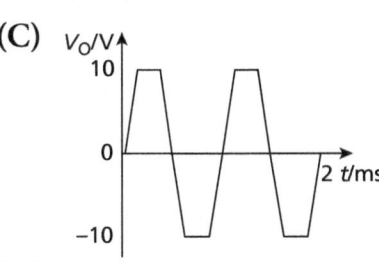

(B)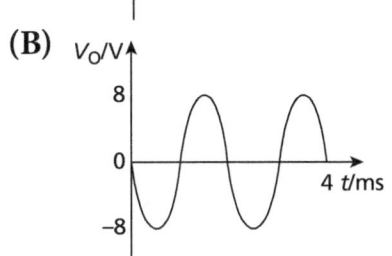

(D)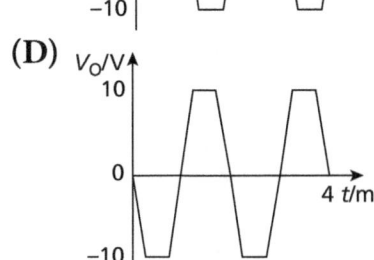

145

2.2.3: Operational Amplifiers (cont.)

21 What is the value of V_O obtained from the system of cascading amplifiers shown?

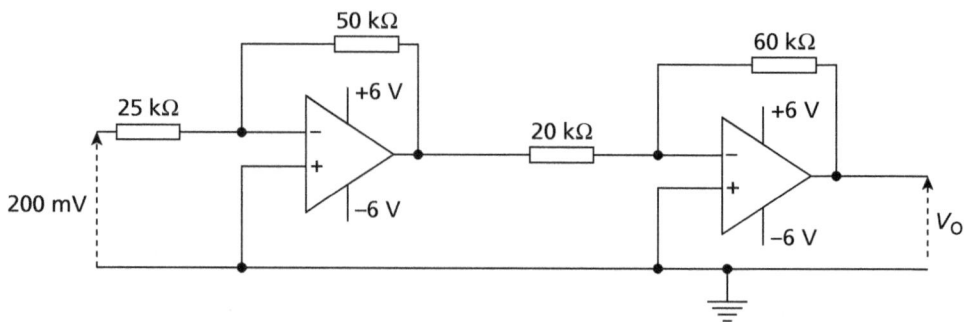

(A) 1.2 V

(B) 1.0 V

(C) −1.2 V

(D) −1.0 V

22 What are the values of the gain and the output voltage of the following amplifier circuit?

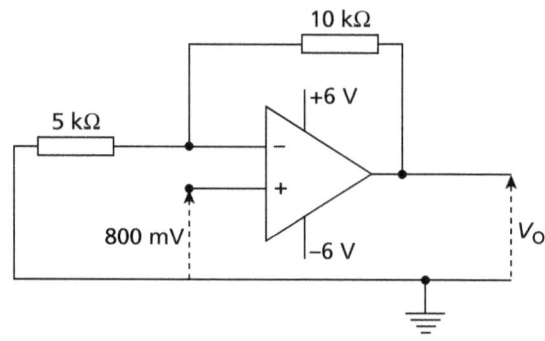

	Gain	V_o/V
(A)	−2	−1.6
(B)	3	2.4
(C)	2	1.6
(D)	−3	−2.4

Items <u>23–24</u> refer to the following graph, which shows the variation of output voltage V_o to input voltage V_i of an amplifier having just one feedback resistor.

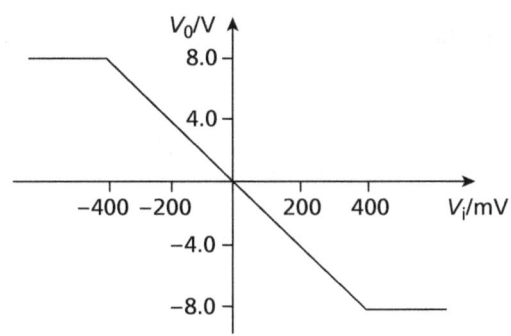

23 What type of amplifier is it and what is its gain?

	Type	Gain
(A)	Inverting	−0.02
(B)	Non-inverting	21
(C)	Inverting	−20
(D)	Voltage follower	1.02

Ⓐ
Ⓑ
Ⓒ
Ⓓ

24 What is the maximum input voltage swing that can occur without the output being saturated?

(A) ±8.0 V

(B) ±400 mV

(C) 16 V

(D) 8 mV

Ⓐ
Ⓑ
Ⓒ
Ⓓ

2.2.3: Operational Amplifiers (cont.)

Item **25** refers to the following diagram.

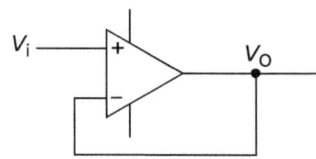

25 Which of the following is NOT true of the circuit?

(A) It is called a voltage follower and is often referred to as a buffer.

(B) It is a special type of non-inverting amplifier having a gain of 1.

(C) It cannot be used in a circuit to measure the e.m.f. of a cell.

(D) It takes no energy from the source connected to its input.

Items **26–27** refer to the following graph, which shows the gain–frequency curve of a practical operational amplifier.

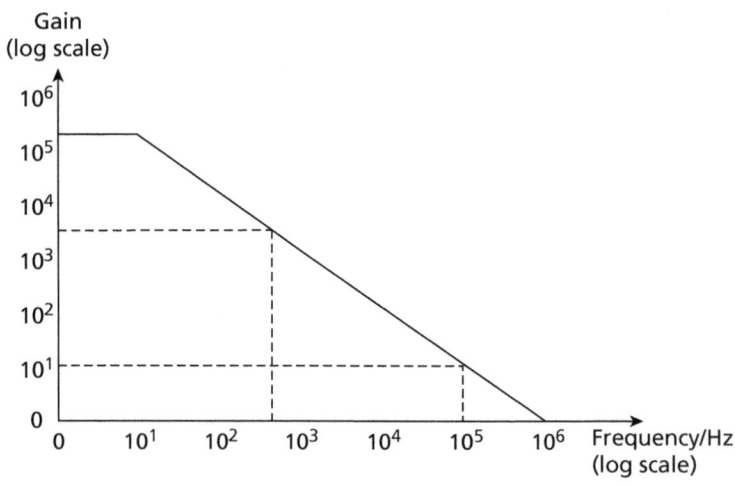

26 Which of the following is the BEST approximation of the bandwidth corresponding to a gain of 3000?

(A) $0 \to 100\,\text{Hz}$

(B) $0 \to 400\,\text{Hz}$

(C) $0 \to 5000\,\text{Hz}$

(D) $0 \to 10\,000\,\text{Hz}$

27 An inverting amplifier uses the operational amplifier in its circuitry to obtain a bandwidth of $0 \to 10^5$. What is the approximate ratio of the feedback resistance to the input resistance?

(A) 10:1

(B) 1:10

(C) 5:1

(D) 1:5

28 Which of the following is NOT true?

(A) Negative feedback reduces distortion.

(B) Non-inverting amplifiers ideally have extremely high input impedance. This is not the case for inverting amplifiers.

(C) Non-inverting amplifiers are useful for measuring the small e.m.f. of thermocouples since they amplify the voltage without taking energy from the source.

(D) Inverting amplifiers use negative feedback and non-inverting amplifiers use positive feedback.

2.2.4: Logic Gates

Items **1–3** refer to the following circuits.

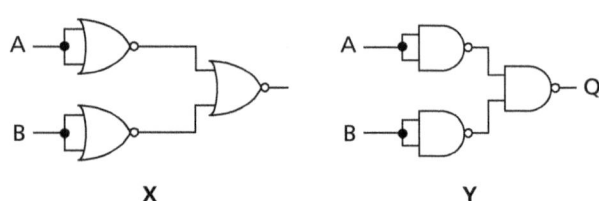
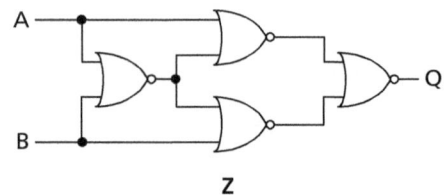

1 What single logic gate is equivalent to X?

(A) AND Ⓐ
(B) OR Ⓑ
(C) NAND Ⓒ
(D) NOT Ⓓ

2 What single logic gate is equivalent Y?

(A) AND Ⓐ
(B) XOR Ⓑ
(C) OR Ⓒ
(D) NOT Ⓓ

3 What single logic gate is equivalent to Z?

(A) NAND Ⓐ
(B) XOR Ⓑ
(C) XNOR Ⓒ
(D) NOR Ⓓ

4 Which of the following sets gives the correct output from EACH of the gates shown?

	X	Y	Z
(A)	1	1	1
(B)	1	1	0
(C)	0	1	0
(D)	1	0	0

Items **5–6** refer to the following diagrams.

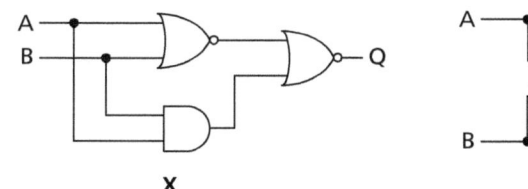
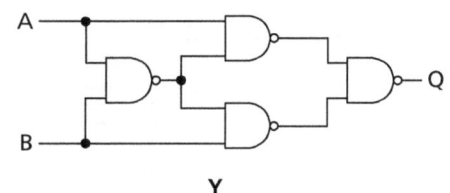

5 What single logic gate is equivalent to X?

(A) NAND

(B) XOR

(C) XNOR

(D) NOR

6 What single logic gate is equivalent to Y?

(A) NAND

(B) XOR

(C) XNOR

(D) NOR

2.2.4: Logic Gates (cont.)

7 What single logic gate is equivalent to the combination of gates shown?

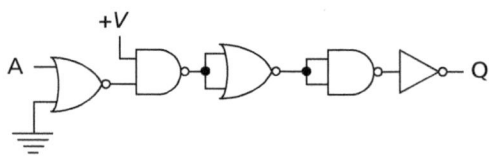

- (A) XOR
- (B) XNOR
- (C) NAND
- (D) NOT

Item 8 refers to the following diagram.

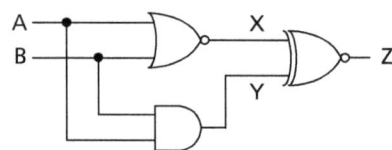

8 If the logic states of A and B are both 1, what are the states at X, Y and Z?

	X	Y	Z
(A)	0	1	0
(B)	0	1	1
(C)	1	1	1
(D)	0	0	1

Item 9 refers to the following diagram, which shows two 8-bit pulse trains X and Y. The pulses are input to a logic gate and exit the gate as a single pulse train Q.

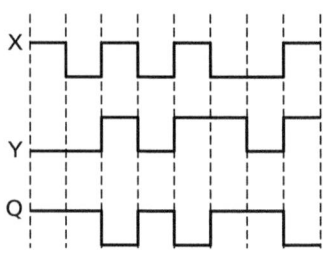

152

9 What is the logic gate to which X and Y are input?

(A) NOR
(B) AND
(C) NAND
(D) OR

10 Which of the rows in the table correctly shows the number of inputs and outputs present in a half-adder and a full-adder?

	Half-adder		Full-adder	
	Inputs	Outputs	Inputs	Outputs
(A)	2	2	2	2
(B)	3	2	3	2
(C)	2	2	3	2
(D)	1	2	3	2

11 Which of the following circuits CANNOT be used as a half-adder?

(A)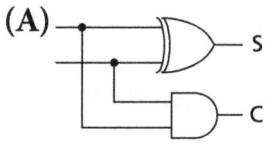

(B)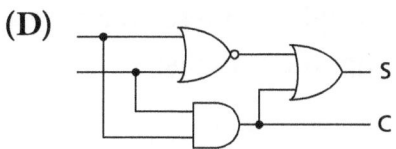

(C)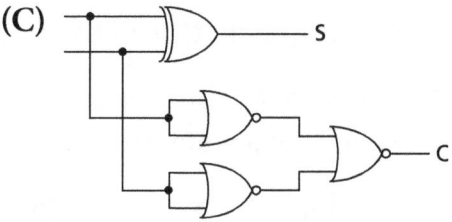

(D)

2.2.4: Logic Gates (cont.)

<u>Item 12</u> refers to the following diagram, which shows two T-type bistables used to light lamps X and Y. The bistables are triggered by a falling edge of a pulse. A graph of voltage V against time t is shown for the clock pulse.

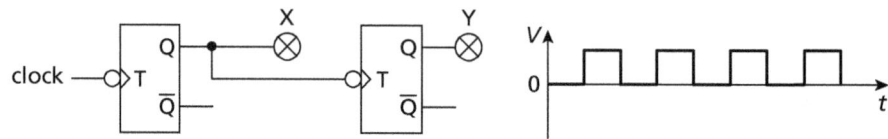

12 For the FOUR input clock pulses shown, the sequence of outputs at X and Y indicating when the lamps are lit (logic 1) is

(A)
X	Y
0	0
0	1
1	0
1	1

(B)
X	Y
0	0
1	0
0	1
1	1

(C)
X	Y
0	0
1	0
1	1
0	1

(D)
X	Y
0	0
0	1
1	1
0	1

Ⓐ Ⓑ Ⓒ Ⓓ

13 Which of the following is/are true of the S-R NOR flip-flop circuit?

  I. The basic S-R NOR flip-flop having just two NOR gates is triggered by the 'high' (logic 1) of a voltage pulse.

  II. When both input lines are 'low' (logic 0), the output state remains latched.

  III. If a 'high' (logic 1) enters on any one of the two input lines, the output state of the flip-flop must change.

(A) I only

(B) I and II only

(C) II and III only

(D) I, II and III

Ⓐ Ⓑ Ⓒ Ⓓ

14 Which of the following is/are true of the basic S-R NOR flip-flop and NAND flip-flop comprised of just two logic gates?

    I. Both are triggered by a 'high' (logic 1) voltage pulse.

    II. Both can act as memory devices.

    III. The NOR flip-flop can have two stable output states for a logic input of 0,0 and the NAND flip-flop can have two stable output states for a logic input of 1,1.

(A) I and II only

(B) II only

(C) I and III only

(D) II and III only

Item 15 refers to the following circuit diagram.

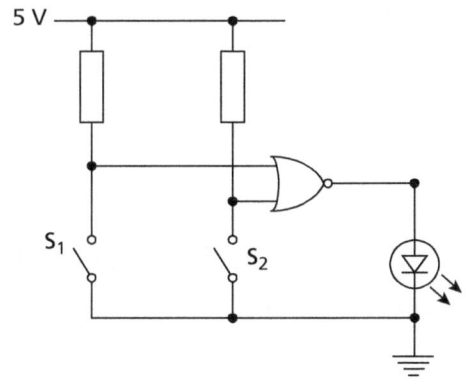

15 What arrangement of the switches is needed for the LED to be lit?

	S_1	S_2
(A)	closed	closed
(B)	open	closed
(C)	closed	open
(D)	open	open

2.2.4: Logic Gates (cont.)

Item 16 refers to the following diagram. The lamps are lit when the inputs to them are logic 1.

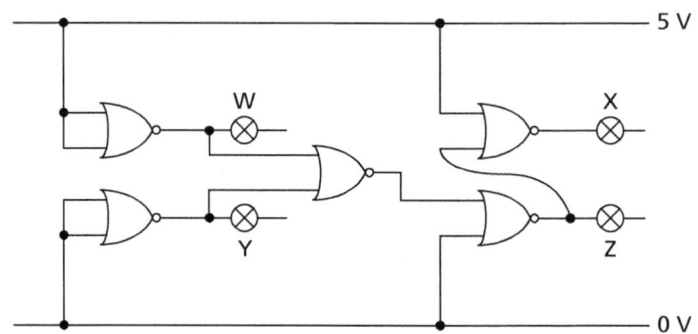

16 Which lamps will be lit?

(A) W and X Ⓐ

(B) Y and X Ⓑ

(C) Y and Z Ⓒ

(D) Y only Ⓓ

Item 17 refers to the flip-flop circuit below. The inputs applied to X and Y in successive periods are shown in the accompanying table.

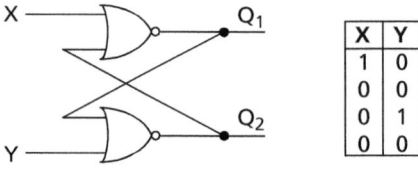

17 What are the corresponding output states, Q_1 and Q_2?

(A)		(B)		(C)		(D)	
Q_1	Q_2	Q_1	Q_2	Q_1	Q_2	Q_1	Q_2
1	0	0	1	1	0	1	0
0	1	0	1	0	1	1	0
0	1	1	0	1	0	0	1
1	0	1	0	0	1	0	1

Ⓐ
Ⓑ
Ⓒ
Ⓓ

Module 3: Atomic and Nuclear Physics
2.3.1: Particulate Nature of Electromagnetic Radiation

Items 1–2 refer to the following diagram. UV light is incident on the photocathode and produces electrons, which are accelerated through the p.d. provided. The direction of the p.d. may be reversed.

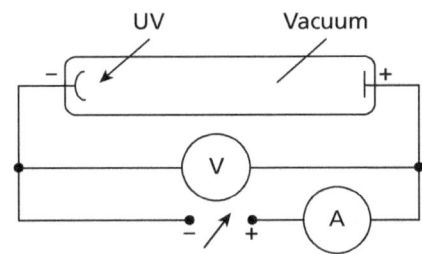

1 Which graph BEST shows how the current I varies with the applied p.d. V?

(A)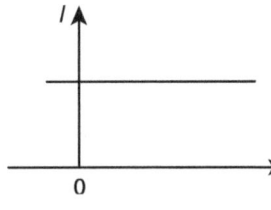

(C)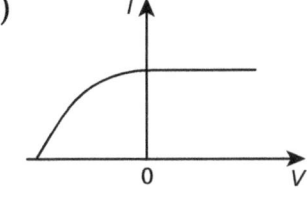

(B)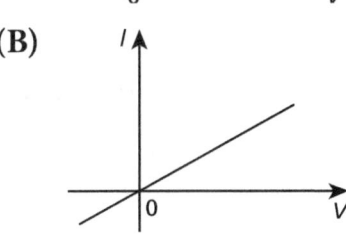

(D)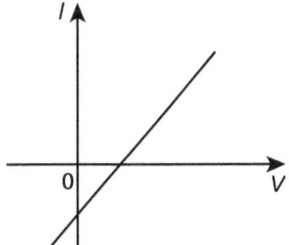

Ⓐ
Ⓑ
Ⓒ
Ⓓ

2 For a constant p.d. which accelerates the electrons, which graph BEST shows how the photoelectric current I will vary with the intensity of the incident radiation?

(A)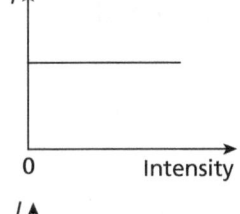

(C)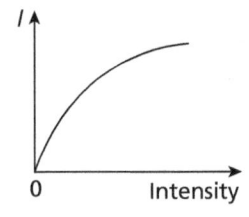

(B)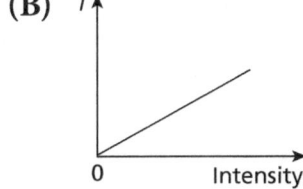

(D)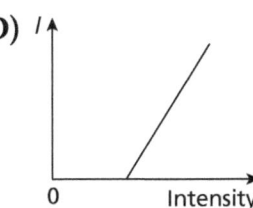

Ⓐ
Ⓑ
Ⓒ
Ⓓ

2.3.1: Particulate Nature of Electromagnetic Radiation (cont.)

Items **3–7** refer to the following graph, which shows the variation in the maximum kinetic energy of photoelectrons emitted by a certain metal with the frequency of the incident radiation.

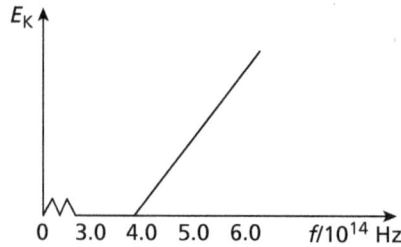

3 What is the work function of the metal?

(A) 1.7×10^{-48} J

(B) 4.0×10^{14} J

(C) 2.7×10^{-19} J

(D) 8.8×10^{-28} J

4 What is the maximum kinetic energy of the photoelectrons emitted when the incident radiation is of frequency 6.0×10^{14} Hz?

(A) 1.3×10^{-19} J

(B) 4.0×10^{-19} J

(C) 6.63×10^{-19} J

(D) 2.7×10^{-19} J

5 What is the cut-off wavelength for photoelectric emission from this metal?

(A) 11×10^{-7} m

(B) 7.5×10^{-7} m

(C) 4×10^{-14} nm

(D) 7.5 nm

6 If h is the Planck constant and the mass and charge of an electron are m and e respectively, then the gradient of the graph represents

(A) $\dfrac{mh}{e}$ (C) h

(B) $\dfrac{m}{e}$ (D) $\dfrac{e}{m}$

7 What is the value of the stopping potential that can prevent emission of photoelectrons when the frequency of the incident radiation is 9.0×10^{14} Hz?

(A) 3.7 V

(B) 1.7 V

(C) 21 V

(D) 2.1 V

8 The electron volt is

(A) the charge on an electron at a potential of 1 volt.

(B) the electrical potential of an electron.

(C) the energy of an electron that has been accelerated by a p.d. of 1 volt.

(D) the field strength which pulls electrons through a p.d. of 1 volt.

9 An energy of 6.4×10^{-20} J can be expressed as

(A) 0.40 eV

(B) 4.0×10^{-20} eV

(C) 0.1 eV

(D) 1.0×10^{-38} eV

10 What is the threshold frequency of a metal if its work function is 4.0 eV and a wavelength of 700 nm is incident on it?

(A) 2.7×10^{-4} Hz

(B) 9.7×10^{14} Hz

(C) 2.5×10^{19} Hz

(D) 4.7×10^{26} Hz

2.3.1: Particulate Nature of Electromagnetic Radiation (cont.)

11 What is the maximum kinetic energy of the photo electrons emitted when radiation of wavelength 300 nm is incident on a metal that has a work function of 2.0 eV?

(A) 2.5 mJ

(B) 4.0×10^{-26} J

(C) 3.4×10^{-19} J

(D) 8.3×10^{-7} J

12 The scientist who explained the photoelectric effect in terms of quantum theory was

(A) Albert Einstein

(B) Max Planck

(C) Neils Bohr

(D) James Clerk Maxwell

13 How does the work function of a metal change if the intensity and the frequency of the radiation incident on it are both doubled?

(A) It quadruples.

(B) It doubles.

(C) It is unaffected.

(D) It is reduced to one quarter of its original value.

14 Which of the following is NOT true of the photoelectric effect?

(A) In order for photoelectric emission to occur, the incident radiation must have a wavelength below a certain value.

(B) If the incident radiation is above the threshold frequency, photoelectric emission will occur instantly.

(C) Increasing the intensity of the radiation increases the speed of the electrons emitted.

(D) The kinetic energy of the electrons emitted depends on the wavelength of the incident radiation.

Items **15–16** refer to the following information and graph.

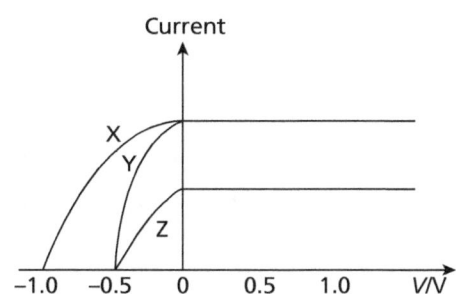

Light of different intensities and frequencies is incident on a metal and the photoelectrons produced are subjected to a variable potential gradient. The graph shows the variation of current and potential difference for radiations X, Y and Z.

15 Radiation X is of frequency 8.0×10^{14} Hz. What is the work function of the metal?

(A) 5.3×10^{-19} J

(B) 3.7×10^{-19} J

(C) 1.6×10^{-19} J

(D) 6.9×10^{-19} J

16 Which of the following is NOT true of the metal and the radiations X, Y and Z?

(A) X is of higher frequency than Y and Z.

(B) The intensity of Y is greater than that of Z.

(C) Y and Z have different wavelengths.

(D) At a forward p.d. of 1.0 V, the speed of the electrons produced by Y is the same as the speed of the electrons produced by Z.

17 Electrons accelerated through a certain potential difference and then stopped by atoms of a gas cause the emission of light of wavelength 500 nm. What is the p.d.?

(A) 2.5 V

(B) 0.40 V

(C) 25 V

(D) 4.0 V

2.3.1: Particulate Nature of Electromagnetic Radiation (cont.)

18 A beam of electromagnetic radiation of wavelength 600 nm and power 0.12 W is emitted from a monochromatic source. What is the number of photons emitted in each second?

(A) 1.1×10^{16}

(B) 2.4×10^{-16}

(C) 3.6×10^{19}

(D) 3.6×10^{17}

Ⓐ
Ⓑ
Ⓒ
Ⓓ

19 What is the de Broglie wavelength of an electron travelling at $\frac{1}{3}$ the speed of light?

(A) 2.4×10^{-12} m

(B) 7.3×10^{-12} m

(C) 6.63×10^{-42} m

(D) 6.63×10^{-26} m

Ⓐ
Ⓑ
Ⓒ
Ⓓ

20 In an experiment, electrons were accelerated towards a thin film of carbon in an evacuated chamber through a p.d. of about 4 kV. They penetrated the film and formed a series of concentric rings on a fluorescent screen.

Which of the following is/are true concerning the results of the experiment?

    I. The experiment reveals the wave nature of particles.

    II. Decreasing the accelerating p.d. brought the rings closer together.

    III. The rings are the result of diffraction.

(A) I only

(B) I and II only

(C) I and III only

(D) II and III only

Ⓐ
Ⓑ
Ⓒ
Ⓓ

Item 21 refers to the following graph showing the variation of stopping potential V_s with frequency f of radiation producing photoelectrons.

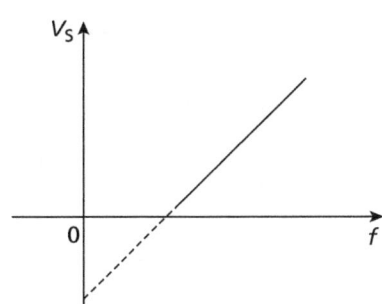

21 If the charge on the electron is e, the work function of the metal is W_0 and the Planck constant is h, then the gradient and the y-intercept of the graph represent:

	Gradient	**y-intercept**
(A)	h	$-W_0$
(B)	$\dfrac{h}{e}$	$\dfrac{-W_0}{e}$
(C)	$\dfrac{h}{W_0}$	$\dfrac{-W_0}{he}$
(D)	h	W_0

22 Which graph correctly shows the relationship between the velocity v of a particle and its de Broglie wavelength λ?

(A)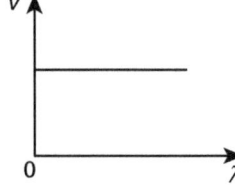

(B)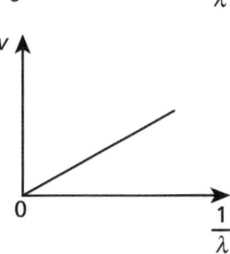

(C)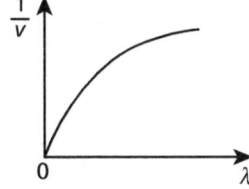

(D)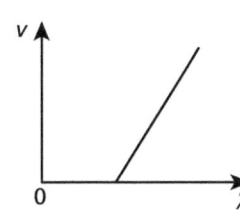

2.3.1: Particulate Nature of Electromagnetic Radiation (cont.)

23 When X-rays pass through 15 cm of a certain material the intensity falls to ¼ of its initial value. What is the linear absorption coefficient of the material?

(A) 9.2 m^{-1}

(B) 1.7 m^{-1}

(C) 0.13 m^{-1}

(D) 3.8 m^{-1}

24 The linear absorption coefficient of a material is 50 m^{-1}. X-rays produce a count-rate of 400 s^{-1} before entering the material and of 100 s^{-1} as they exit from the side directly opposite. What is the thickness of the material?

(A) 0.25 cm

(B) 1.4 cm

(C) 0.50 cm

(D) 2.8 cm

25 What is the de Broglie wavelength of a particle of mass m and of kinetic energy E?

(A) $\dfrac{2h}{mE}$

(B) $\dfrac{h}{2mE}$

(C) $\dfrac{h}{(2mE)^{\frac{1}{2}}}$

(D) $\dfrac{(2mE)^{\frac{1}{2}}}{h}$

Items **26–28** refer to the following diagram, which shows various energy levels of the hydrogen atom. The quantum number n and the energy E of particular levels are given.

```
n = ∞  ----------------------------------  E∞ = 0
n = 5  ──────────────────────────────────  E₅ = –0.54 eV
n = 4  ──────────────────────────────────  E₄ = –0.85 eV
n = 3  ──────────────────────────────────  E₃ = –1.5 eV

n = 2  ──────────────────────────────────  E₂ = –3.4 eV


n = 1  ──────────────────────────────────  E₁ = –13.6 eV
```

26 A photon of frequency 3.08×10^{15} Hz is emitted when an electron jumps to a lower energy level. The transition is

(A) $E_4 \to E_2$

(B) $E_3 \to E_1$

(C) $E_3 \to E_2$

(D) $E_4 \to E_1$

Ⓐ
Ⓑ
Ⓒ
Ⓓ

27 A fast-moving electron is accelerated through a p.d. and bombards a hydrogen atom, causing an orbital electron to jump from $n = 1$ to $n = 3$. What is the value of this p.d.?

(A) 1.5 eV

(B) 1.5 V

(C) 12.1 V

(D) 12.1 eV

Ⓐ
Ⓑ
Ⓒ
Ⓓ

2.3.1: Particulate Nature of Electromagnetic Radiation (cont.)

28 What is the ionisation energy of a hydrogen atom in its ground state?

(A) 3.4 eV

(B) 2.2×10^{-18} J

(C) 13.6 J

(D) 13.6×10^{-19} eV

29 An electron makes a transition from E_3 to E_1, producing a photon of wavelength λ in the process. Which of the following expresses Planck's constant in terms of this data?

(A) $\dfrac{(E_1 - E_3)\lambda}{c}$

(B) $\dfrac{(E_1 - E_3)c}{\lambda}$

(C) $\dfrac{(E_3 - E_1)\lambda}{c}$

(D) $\dfrac{(E_3 - E_1)c}{\lambda}$

30 Which of the following graphs relating the energy, frequency and wavelength of an electromagnetic wave is INCORRECT?

(A)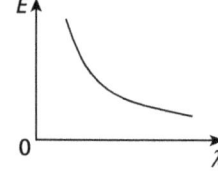

(B)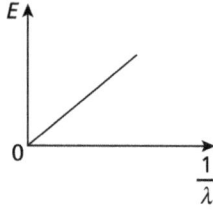

(C)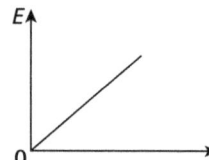

(D)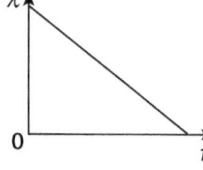

Items **31–33** refer to the following graph, which shows an X-ray spectrum.

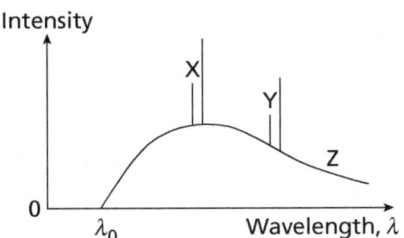

31 Which of the following is/are true?

   I. X and Y are line spectra which are produced due to electron transitions between energy levels. They are therefore characteristic of the element emitting the X-rays.

   II. λ_0 is the wavelength corresponding to the least energetic X-ray.

   III. Z is the continuous or white spectrum produced as bombarding electrons decelerate within the target and come to rest during X-ray production.

   (A) I and II only
   (B) I and III only
   (C) II and III only
   (D) III only

32 What is the value of λ_0 if the X-ray tube has an accelerating p.d. of 120 kV?

   (A) 1.0×10^{-11} m
   (B) 5.2×10^{-20} m
   (C) 2.1×10^{-12} m
   (D) 4.3×10^{-13} m

33 What is the maximum velocity of the electrons bombarding the target when the p.d. is 120 kV?

   (A) 1.9×10^{7} m s^{-1}
   (B) 3.2×10^{7} m s^{-1}
   (C) 2.1×10^{8} m s^{-1}
   (D) 2.5×10^{7} m s^{-1}

2.3.1: Particulate Nature of Electromagnetic Radiation (cont.)

34. X and Y are photons with energies of 12.0 eV and 8.0 eV respectively. Which of the following is NOT true of X and Y?

(A) X has a smaller wavelength.

(B) Y has energy of 1.28×10^{-18} J.

(C) X is more energetic and therefore travels at a greater speed.

(D) Y has a larger wavelength than X.

2.3.2: Atomic Structure and Binding Energy

1. The Geiger–Marsden experiment revealed valuable information with respect to

(A) the various energy levels within the atom.

(B) the instability of the nucleus.

(C) the nature and position of the neutrons in the atom.

(D) the small concentrated positive nucleus of an atom.

2. Uranium has an atomic number of 92 and a certain isotope of it has a mass number of 235. Which of the following is true of a neutral atom of this isotope of uranium?

	Number of nucleons	Number of protons	Number of neutrons	Number of electrons
(A)	235	92	143	143
(B)	143	92	143	92
(C)	235	143	92	143
(D)	235	92	143	92

3 Which of the following is NOT true of two neutral atoms of different isotopes of the same element?

(A) Each atom has the same atomic number but contains different numbers of neutrons.

(B) In each atom the number of protons is equal to the number of electrons.

(C) The number of protons each contains may be obtained by subtracting the number of nucleons from its mass number.

(D) The atomic number of each may be obtained by subtracting the number of neutrons it contains from its nucleon number.

4 In a Millikan's oil drop experiment, an oil drop of density ρ and radius r is balanced between two horizontal plates which are a distance d apart and have a p.d. of V between them. If the charge on the drop is q, what is the value of V?

(A) $\dfrac{3\pi r^3 \rho g d}{4q}$

(B) $\dfrac{4\pi r^3 \rho g d}{3q}$

(C) $\dfrac{4gd\rho}{3q\pi r^3}$

(D) $\dfrac{4qgd}{3\pi r^{3\rho}}$

5 In a Millikan's oil drop experiment, a drop of mass m and charge q is balanced by a p.d. of V. What p.d. is required to balance the same drop if the charge on it is $4q$?

(A) $\dfrac{4Vm}{q}$

(B) $\dfrac{4q}{mV}$

(C) $\dfrac{V}{4}$

(D) $4V$

6 In a Millikan's oil drop experiment, a drop of mass m and charge q is balanced by a p.d. of V. What p.d. is required to balance another drop if its mass is $2m$ and its charge is $2q$?

(A) $\dfrac{4Vm}{q}$

(B) V

(C) $4V$

(D) $\dfrac{Vm}{4q}$

2.3.2: Atomic Structure and Binding Energy (cont.)

7 During a Millikan's oil drop experiment, which of the following is NOT true?

(A) Neglecting buoyancy forces, if a drop moves upward the electrical force can be the same in magnitude as the gravitational force.

(B) If the drop is at rest, it must be charged.

(C) If the charge of a balanced drop is doubled but its mass remains the same, then the p.d. to maintain balance must also be doubled.

(D) If the mass of a balanced drop is doubled but the charge on it remains the same, then the p.d. to maintain balance must also be doubled.

8 Which of the following is NOT true of a Millikan's oil drop experiment?

(A) It reveals that charge is quantised.

(B) It reveals that there is a minimum charge that a body can acquire.

(C) It reveals that there is a constant charge-to-mass ratio for all oil drops.

(D) During the experiment, a greater p.d. is needed to balance a drop if the charge on it is reduced.

9 During a Millikan's oil drop experiment, several charges Q were investigated and it was found that their magnitudes were in the following relative proportions.

Q_1 Q_2 Q_3 Q_4 Q_5 Q_6 Q_7

5.0   7.5   10.0   12.5   15.0   17.5   20.0

According to Millikan, what is the relative proportion of the smallest possible charge using this data?

(A) 2.5

(B) 8.0

(C) 10.0

(D) 20.0

10 Which of the following is/are true?

  I. Binding energy is the energy released when an atomic nucleus is assembled from its nucleons at infinity.

  II. Atoms of atomic number greater than that of iron do not release a net amount of energy when they undergo nuclear fusion.

  III. The binding energy of a uranium atom is the energy it releases on its fission reaction as it converts to barium and krypton.

  IV. The mass of an atomic nucleus is greater than the mass of its constituent nucleons when infinitely separated.

(A) I and II only

(B) II only

(C) II and III only

(D) I and IV only

11 The binding energy per nucleon of the nucleus of an atom of $^{4}_{2}\text{He}$ is 7.05 MeV. Which of the following correctly shows the loss in mass in kilograms and in atomic mass units on formation of a nucleus of $^{4}_{2}\text{He}$?

	kg	u
(A)	2.0×10^{-29}	0.0076
(B)	5.0×10^{-29}	0.0303
(C)	2.0×10^{-28}	0.0303
(D)	5.0×10^{-29}	0.0076

12 The mass of the nucleus of the atom $^{4}_{2}\text{He}$ is 4.0015 u and the mass of its constituent particles separated to infinity is 4.0319 u. What is the binding energy of the nucleus?

(A) 2.74×10^{15} J

(B) 1.51×10^{-20} J

(C) 4.54×10^{-12} J

(D) 5.98×10^{-10} J

2.3.2: Atomic Structure and Binding Energy (cont.)

13 6.41 MeV of energy is released in a nuclear reaction. The mass loss in the process is

(A) 0.002 59 u

(B) 0.006 89 u

(C) 0.007 12 u

(D) 0.005 42 u

14 What is the energy corresponding to the atomic mass unit?

(A) 931×10^6 J

(B) 9.0×10^{16} J

(C) 1.49×10^{-10} J

(D) 4.98×10^{-19} J

15 The decay of thorium is as follows

$$^{228}_{90}\text{Th} \rightarrow {}^{224}_{88}\text{Ra} + {}^{4}_{2}\text{He} + \text{energy}$$

Mass of $^{228}_{90}$Th = 228.02873 u

Mass of $^{224}_{88}$Ra = 224.02020 u

Mass of $^{4}_{2}$He = 4.00260 u

What mass is lost in the decay?

(A) 6.59×10^{-20} kg

(B) 4.25×10^{-30} kg

(C) 9.84×10^{-30} kg

(D) 1.98×10^{-11} kg

Item **16** refers to the following graph.

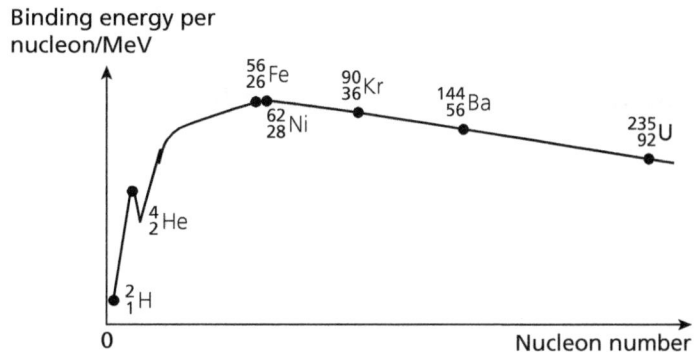

16 Which of the following is NOT true?

(A) The nuclei of $^{56}_{26}$Fe and $^{62}_{28}$Ni are the most stable.

(B) Elements to the left of $^{56}_{26}$Fe are more likely to undergo fission than those to the right of it.

(C) The potential energies of all the elements are negative.

(D) Fusion reactions require temperatures of over 1×10^8 K.

17 What are the values of *w, x, y* and *z* in the following nuclear equation?
$^{235}_{92}U + ^{w}_{x}n \rightarrow ^{y}_{56}Ba + ^{92}_{z}Kr + 3^{w}_{x}n$

	w	x	y	z
(A)	0	1	141	36
(B)	1	0	144	90
(C)	0	1	144	90
(D)	1	0	141	36

2.3.3: Radioactivity

1 A radioisotope decays to $\frac{1}{8}$ of its mass in 24 days. What is its half-life?

(A) 3 days

(B) 6 days

(C) 8 days

(D) 12 days

2 What is the decay constant of a radioisotope if 2.0 kg of it has a half-life of 5700 years?

(A) 8.23×10^3 per year

(B) 1.2×10^{-4} per year

(C) 0.6×10^{-4} per year

(D) 4.1×10^{-4} per year

3 How long will it take for a radioisotope to decay from 3.0×10^6 particles to 7.5×10^5 particles if its decay constant is 1.0×10^{-2} per year?

(A) 29

(B) 69

(C) 119 years

(D) 140 years

4 What percentage of a radioisotope of half–life 6 hours remains after a period of 30 hours?

(A) 3 %

(B) 6 %

(C) 13 %

(D) 20 %

5 The count rate obtained when 1.0 g of carbon from a living plant is placed in front of a Geiger detector is 25 per minute. 10 g of carbon extracted from an old wooden weapon when placed at the same distance in front of the detector indicates a count rate of 17.5 per minute. What is the age of the old weapon if the background count rate is 5 per minute and the half-life of carbon-14 is 5700 years?

(A) 4000 years

(B) 11 400 years

(C) 17 100 years

(D) 23 000 years

Ⓐ
Ⓑ
Ⓒ
Ⓓ

6 The radioisotope $^{216}_{84}$Po decays by a series of emissions to a stable isotope. Two alpha particles and two beta particles are emitted in the process. Which of the following is the resulting nuclide?

(A) $^{212}_{82}$Pb

(B) $^{208}_{82}$Pb

(C) $^{208}_{83}$Bi

(D) $^{208}_{81}$Tl

Ⓐ
Ⓑ
Ⓒ
Ⓓ

7 A thorium nucleus decays by a series of two α-particle and two β-particle emissions to an isotope, $^{224}_{88}$Ra. How many protons exist in a nucleus of this isotope of thorium?

(A) 89

(B) 90

(C) 92

(D) 94

Ⓐ
Ⓑ
Ⓒ
Ⓓ

2.3.3: Radioactivity (cont.)

8 The decay of $^{220}_{86}Rn$ to $^{208}_{82}Pb$ is part of a radioactive decay series. Which of the following emissions are possible for this decay?

(A) 3 α-particles and 2 β-particles and 1 γ-ray

(B) 4 α-particles and 3 β-particles and 2 γ-rays

(C) 4 α-particles and 2 β-particles and 1 γ-ray

(D) 3 α-particles and 3 β-particles

9 Which of the following are true of radioactive decay?

　　I. It is dependent on physical and chemical conditions.

　　II. It is a result of instability within the atomic nucleus.

　　III. It is a random and spontaneous phenomenon.

　　IV. Unlike fission and fusion, there is no output of energy during the process.

(A) I, II and III

(B) II and III only

(C) II, III and IV only

(D) II and IV only

10 Which of the following properties would be most suitable for a radioactive tracer?

(A) Alpha emitter of half-life 4 hours.

(B) Beta emitter of half-life 5 minutes.

(C) Gamma emitter of half-life 6 hours.

(D) Gamma emitter of half-life 40 years.

11 A radioactive source that emits gamma radiation is placed at a distance x in front of a Geiger detector in the laboratory. Which of the following graphs is true of the count rate C_D detected if there is background radiation in the region?

(A)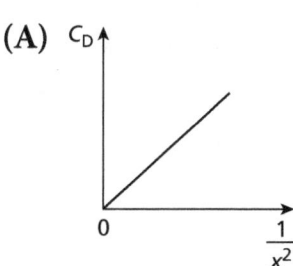

(C)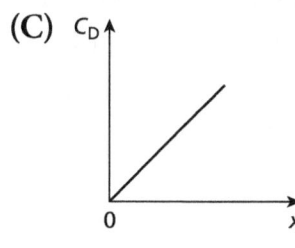

(B)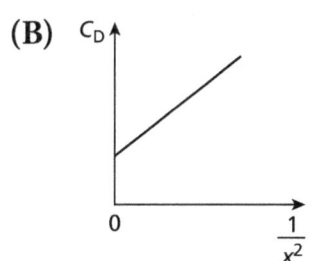

(D)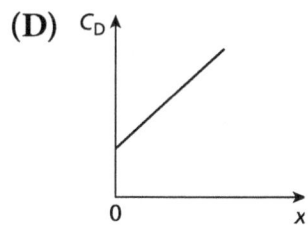

Ⓐ Ⓑ Ⓒ Ⓓ

12 Which of the following is true of radioactive decay?

(A) The activity of a source is the count rate detected by a Geiger detector placed in front of it.

(B) The count rate obtained from a source decreases uniformly with time.

(C) Half-life is half the time for a sample of a radioactive material to decay.

(D) A few isotopes can emit α-particles and β-particles.

Ⓐ Ⓑ Ⓒ Ⓓ

2.3.3: Radioactivity (cont.)

13 The following diagram shows a source of bismuth-212, which can emit α and β radiation. Detectors were positioned at Q, R, S and T, and both radiations were detected. At which point is the detection of each most likely if a magnetic field acts perpendicularly into the plane of the paper on which the diagram is drawn?

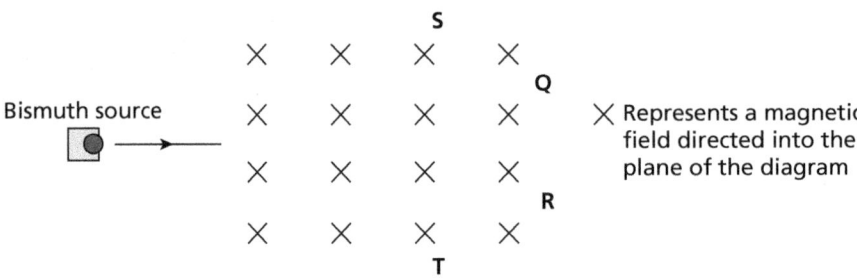

	α-detection	β-detection
(A)	T	Q
(B)	R	S
(C)	Q	T
(D)	S	R

14 Which of the following is NOT true of the Geiger–Müller tube?

(A) A p.d. of about 400 V is suitable between the central anode rod and the case of the detector.

(B) It typically contains argon gas at a low pressure.

(C) The window at the front is very thin and is made of mica.

(D) Bromine is added as a quenching agent to the chamber, to absorb electrons produced by collision of the positive ions with the cathode and thereby to increase the time for which the discharge takes place.